Data Visualization

Charts, Maps and Interactive Graphics

ASA-CRC Series on
STATISTICAL REASONING IN SCIENCE AND SOCIETY

SERIES EDITORS

PUBLISHED TITLES

Errors, Blunders, and Lies: How to Tell the Difference
David S. Salsburg

Visualizing Baseball
Jim Albert

Data Visualization: Charts, Maps and Interactive Graphics
Robert Grant

For more information about this series, please visit:
https://www.crcpress.com/go/
https://www.crcpress.com/go/asacrc

Data Visualization

Charts, Maps and Interactive Graphics

Robert Grant

CRC Press is an imprint of the
Taylor & Francis Group, an **informa** business
A CHAPMAN & HALL BOOK

Cover image: Landscape of Change uses data about sea level rise, glacier volume decline, increasing global temperatures, and the increasing use of fossil fuels. These data lines compose a landscape shaped by the changing climate, a world in which we are now living. Copyright © Jill Pelto (jillpelto.com).

CRC Press
Taylor & Francis Group
6000 Broken Sound Parkway NW, Suite 300 Boca Raton, FL 33487-2742

CRC Press is an imprint of Taylor & Francis Group, an Informa business

Printed on acid-free paper
Version Date: 20181103

International Standard Book Number-13: 978-1-138-70760-3 (Paperback)
International Standard Book Number-13: 978-1-138-55359-0 (Hardback)

Library of Congress Cataloging-in-Publication Data

Names: Grant, Robert, 1974- author.
Title: Data visualization : charts, maps, and interactive graphics / by Robert Grant.
Description: Boca Raton, Florida : CRC Press, [2019] | Includes bibliographical references and index.
Identifiers: LCCN 2018028995| ISBN 9781138707603 (pbk. : alk. paper) | ISBN 9781138553590 (hardback : alk. paper) | ISBN 9781315201351 (e-book).
Subjects: LCSH: Information visualization.
Classification: LCC QA76.9.I52 G73 2019 | DDC 001.4/226--dc23
LC record available at https://lccn.loc.gov/2018028995

Visit the Taylor & Francis Web site at
http://www.taylorandfrancis.com

and the CRC Press Web site at
http://www.crcpress.com

To my parents, who never tired of buying me nerdy books.
I miss you.

Contents

SECTION II Statistical building blocks

List of Figures

Preface

This is the age of data. There are more innovations and more opportunities for interesting work with data than ever before, but there is also an overwhelming amount of quantitative information being published every day. If you consume it (and who doesn't?), you might feel that you want to be able to get more out of what you see. If you produce it, you'll want your message to get across and not fizzle out because of poor communication. I hope this book is helpful to both the producers and the consumers.

Data visualization has become big business, because communication is the difference between success and failure, no matter how clever the analysis may have been. Even the fact that charts and graphs have a new name tells you that it is a hot topic, and you can shorten it to dataviz if you want to be really *à la mode*.

This book is deliberately a broad overview that covers a lot but not in great detail. Part 1 has two chapters that everyone should read, then Part 2 can be skipped if you have already studied statistics, otherwise it will help get you ready for the real fun which follows in Part 3: a series of short chapters of a few pages each, which you can dip into depending on your interests. I know my readers may be busy people, so you can read a Part 3 chapter in a coffee break or commute. You will also find **no algebra** anywhere in this book. I promise.

When we analyze data, we are representing all the complexity of the world, and our attempts to understand how it works, in numbers. In doing so we lose some detail – we choose what to quantify and what to leave behind – but we gain objectivity. With modern statistical methods, we can spot a huge range of patterns amid the noise, and overcome the cognitive biases that confuse us. However, there are limits to what can be seen in a list or even a table of numbers. When the numbers overwhelm us, an image can help us spot what is really of interest. Sometimes, the pattern we are looking for can be complex, and even with only a few numbers, we still can't grasp it until it is drawn. Why do images help so much when they seem fundamentally different from numbers?

One explanation is that we humans have been using our eyes to find food and shelter, and avoid predators, for a very long time, and the part of the brain that processes visual information is extremely well developed. In contrast, we have been writing and reading numbers for only a few thousand years at best. It would be a mistake not to tap into the incredible power of our eyes and brains when we need to understand data.

It can feel like drawing charts and maps belongs to the arts a bit more than the sciences, and somehow is not as respectable as a formidable equation with lots of arcane symbols, or a mind-numbingly large table packed full of numbers. This is really not true: there is nothing less scientific about an image, but there *is* something bad about depicting data in a way that the audience might not understand. If you want to calculate, you should also know how to communicate. I'm sorry to say my profession (statisticians) are not best known for this.

Most of the time, there is no single best way to visualize data, which makes this endeavor harder work but also more interesting. Statistician John Tukey wrote in 1977, "Aims can differ, and plots [visualizations, we might say now] should follow aims." This is still true.

There are several excellent books on data visualization but this one tries to do something different: provide a brief overview of techniques and tools, while all the time emphasizing statistical reasoning – the title of this series. Most of the visualization books of recent years were written by people from a design or journalism background, and although there is much to be learned from them, this book will go deeper into statistical topics like predictive models; I don't believe there has been a statistical overview since William Cleveland's books *The Elements of Graphing Data* in 1985 and *Visualizing Data* in 1993.

I have written with some audiences in mind:

- The manager of a team working with data who has to get them up and running with data visualization, or maybe commission someone to do the work,
- the statistician or data scientist who has never had any training in visualizing their findings and wants to find out about it,

- the designer or journalist who wants to get into working with data in a statistically sound way,
- anyone who reads about data and statistics and wants to know how to get the most information out of charts, graphs and maps while also spotting the misleading or confusing ones.

I created all the visualizations in this book, except where copyright is given in the captions, and on the accompanying website at **robertgrantstats.co.uk/dataviz-book** you will find how I did it in each case. The goal was to illustrate specific topics in dataviz, so they are generally quite plain. I also used parts of datasets, invented datasets, and less than optimal analyses for this reason. You should definitely not rely on any images in this book to guide you on how to make British trains run on time, treat drug addiction, or find taxis in New York!

There are some topics which are fascinating but can't be included in this book: representing data in sound, art made from data, physical representations like sculptures or 3-D printing, visualizing qualitative data (such as interview transcripts) and accessibility for people with visual impairment.

TERMINOLOGY

There are a few common terms that are worth defining before we begin.

- An *observation* is a single set of measurements. When we visualize data, we are dealing with measurements on observations. If it's a questionnaire that was filled out by customers after buying from your business, each questionnaire is an observation. If you have weather data taken at noon every day for several years, each day is an observation.
- A *variable* is a characteristic that we measure for the observations. Your customers might indicate their age – that's a variable – and the weather might include temperature at noon – another variable. Some variables are numbers and some are categories, and we'll separate those in Chapters 3 and 4. Each variable takes a *value* for each observation.
- Each variable is either *quantitative* or *categorical*. The quantitative ones have numbers that mean something, like the

temperature at noon. We can subtract one day's temperature from another and that is a meaningful number: how much warmer or cooler it is. Categorical variables can be sub-divided into *nominal* (just a collection of categories, like countries of the world) or *ordinal* (categories that go in a specific order, like age groups: under 18, 18–30, 31–40, 41–64, 65 and over). These types of variables call for different types of visualization, and I'll go into that in more detail in Part 2 of the book.

- In many data visualizations, there are *axes* which are like rulers alongside the image, and indicate what value is shown at what location. What we're doing is taking the values and encoding them as the position on the page or screen, and the axes show us what that position means in numbers. We'll talk about that more in Chapter 2.
- Sometimes, the data appear as lines, sometimes as colored areas, and sometimes as little circles or other shapes. Those little shapes are called *markers*.

I also want to differentiate between three ways in which the data get translated into what you see:

- By *encoding*, I mean a way of translating numbers into what is seen. Perhaps a high value means that marker is far to the right, and a low value means far to the left.
- By *format*, I mean that there are some familiar looks to visualizations and we can appeal to those to help readers understand what we're showing. The same encoding can result in lines or markers or colored areas, for example – those are formats.
- What remains is a lot of choices about how it looks and I call this *aesthetics*. For example, once you've decided to have a line, you could make it thick or thin, it could have a shiny effect so it looks like a metal tube, or whatever other special effect you want.

If you have encountered Leland Wilkinson's "grammar of graphics" before (which is implemented in the popular "ggplot2" package for the R programming language), you'll notice these are almost the "co-ordinates, geometries and aesthetics" used there (but not quite).

I

The basics

CHAPTER 1

Why visualize?

I SPENT TWENTY YEARS working in various aspects of medical research. I was trained in statistics and contributed to many reports and scientific journal articles. One way or another, my colleagues and I were aiming to make the world a better place. To do that, we first had to get our messages across to busy readers. We needed them to remember and act on our findings.

But most of that effort went on to have no impact at all. Those reports, if they haven't gone to the dump, gather dust on shelves. Those scientific journal articles likewise. Few people have the energy to plough through boring, dry text and tables of numbers. I realized too late that all our data and findings had to be *visualized* to have impact. I thought I knew about it, but I was only producing default bar charts and the like from my software, without thinking through their design and the messages they should convey.

Around 2011, I started really learning how to think critically about data visualization, and was inspired by the amazing examples that were appearing all the time: animations, interactive web pages, and sophisticated statistical stories in newspapers. The term "data visualization" (or dataviz) captured the attention of journalists, academics and business people alike. Why should a picture be *so* powerful compared to a bunch of numbers?

1.1 BECAUSE OUR BRAINS ARE WIRED THAT WAY

When we read some numerical facts in a newspaper article, or see them presented in slides, our brains process this abstract information and try to turn it into a deeper, more intuitive understanding of the patterns. You could just tell people the headline messages, but why should they believe you? If you want to be taken seriously, you should have data to support a message, and you should show your readers that data.

You could do that in words (but it would get extremely repetitive and boring), or you could do it in a table (but it would be enormous and only the most numerate readers will bother to look at it). But if you do it in a visualization, it is compact, accessible to everyone, gets the top message across quickly, and can even be eye-catching and exciting.

Human brains are wired for seeing patterns and differences, and for understanding spatial relationships from this. Just think about how much calculation takes place when you look down the road, see a car approaching in the distance, and decide to cross because you will be on the other side before it arrives. This is all to do with spotting tiny clues about speed and distance, and weighing them up in a fraction of a second. You don't even think about it, it just happens in the more mysterious recesses of your brain that deal with distances, relative positions and changes, and they can distill that information from what your eyes see almost instantaneously.

It takes teams of experts working for years to make a self-driving car that can do the same task. We have evolved to be exceptionally good at dealing with information like that – it was the difference between catching the antelope to feed the family, and being trampled by it – so when we present complex numerical patterns visually, our readers are tapping into all that brain power without even knowing it. It's so effortless that conscious thought is freed up for more important tasks. The reader can focus instead on deciding whether they believe it or whether it matches what they've heard before.

Of course, there are times when you only have a few numbers to get across, and text works well. When the numbers build up and text starts to sound repetitive, a table can be the most sensible option, but – however fascinating you think it is – many readers won't want to read a table of numbers at all, so a visualization can get the message through.

Sometimes the pattern you want them to see is too hard to make out in a table but very clear visually. Tables of data are also useful to back up the visualization with its underlying numbers, so that the really keen readers can look them up. Layering information like this, from the big headline through the more subtle patterns down to the raw data, serves all our readers well. The boss who wants to know the bottom line can just take that from the headline, the

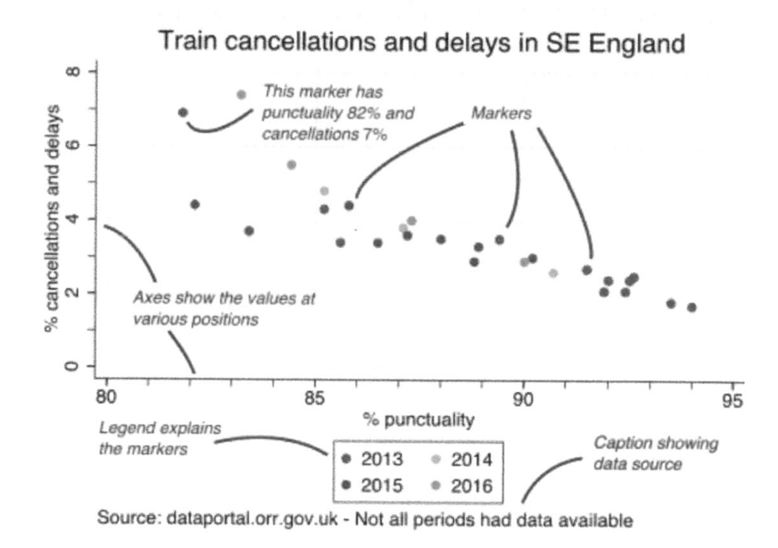

Figure 1.1 How to read a scatter plot.

experts who want to understand the patterns can take that from the visualizations, and the skeptics who want to see exactly how you came up with your conclusions can look at the tables.

If you are going to make data visualizations, take on the role of a curator, explaining and guiding the reader through the information with accompanying text, while leaving them to make their own informed interpretations.

How to read a scatter plot

Any point on the scatter plot can be located exactly by saying how far along the horizontal axis (or "x-axis") it is, and how far up the vertical axis (or "y-axis"). There should be tick marks and numbers along each axis that will help you find each point's values. There could also be "grid lines" that help if precise location matters to the readers. You can find each value by looking at the vertical axis to see how high up a marker is, and what number that relates to, and you can do the same for the horizontal axis. So, this type of plot contains two variables' worth of data.

1.2 TO HELP THE ANALYST AVOID PROBLEMS

Visualizing is not only useful for communicating messages, but also for the analyst to understand their data in depth. Statistics only take us so far in this regard. Let's imagine we have data on two variables. By that, I mean that we have a series of measurements with two values in each. We might be measuring height and weight of people taking part in some study, or the unemployment and homelessness levels in several towns.

For now, I'm just going to imagine an abstract situation where I get the computer to tell me the basic stats that describe these two variables – the numbers that statisticians like me would generate to help them understand the data. It tells me that variable X has a mean (average) of 50.7 and variable Y has mean of 46.5, so the X values are little bit bigger than the Y's. I also learn that the standard deviations are 19.6 and 27.3, so the Y values are a bit more spread out than the X's. I also learn that they have a Pearson correlation of −0.18, which tells me that as values of X get higher, the values of Y tend to get slightly lower. (These terms like standard deviation and correlation are not important for now; we'll get into them in Chapter 5).

Most of the time, analysts stop at that and say they now understand the data. If you asked me to sketch it, I would draw a cloud of dots centered on those mean values, very slightly elongated from top left to bottom right (that's the correlation), and somewhat longer in one direction than the other (the difference in standard deviations). In fact, I would have been fooled, because if I visualize them in a scatter plot (which I'll explain in full in Chapter 2) they look like Figure 1.2!

Admittedly, when you collect real-life data, it is unlikely to look like the Datasaurus. But it could still bite you if you jump straight to statistics without having spotted errors or understood the salient patterns. Visualizations are especially useful for this in the early stages of analysis. It's good practice in data analysis to generate a lot of charts, maps and other relevant visualizations for your own benefit. They will almost all be thrown away but they might help you spot one error or better understand one relationship that saves a lot of time or embarrassment later.

Visualization pioneer William Cleveland wrote, "Graphical methods tend to show data sets as a whole, allowing us to summarize the general behavior and to study detail. This leads to much more thorough data analyses." With statistics alone, you usually

N = 157 ; X mean = 50.7333 ; X SD = 19.5661 ; Y mean = 46.495 ; Y SD = 27.2828 ;
Pearson correlation = -0.1772

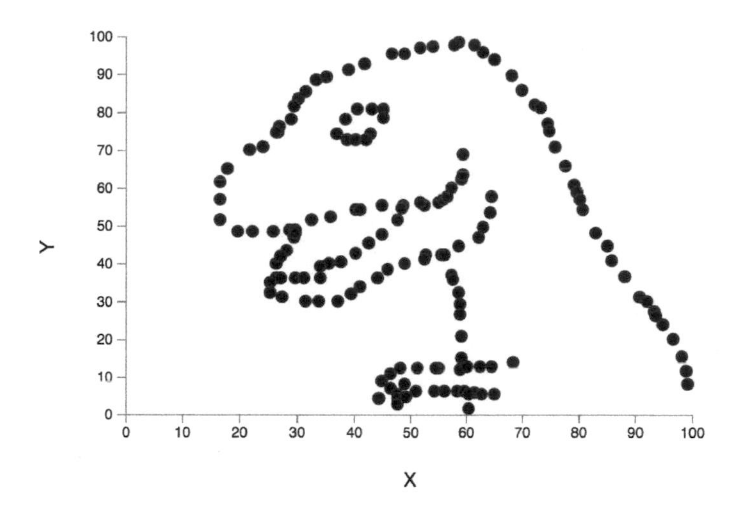

Figure 1.2 The Datasaurus, created by Alberto Cairo using my webpage DrawMyData. Reproduced with permission.

get the general behavior but lose the detail. A classic example of this is the four artificial datasets created by statistician Frank Anscombe, which all have the same means, standard deviations, and correlations (Figure 1.3).

In other words, if you didn't draw the charts, and just worked out the stats, you would conclude that they are identical. Figure 1.3 shows just how different they are. The diagonal line in each chart is the best prediction of where the dots will lie, and for all but the top left chart, it fails to take the appearance of the data into account – but with identical statistics, only drawing the charts will help refine that prediction.

We are also very good at spotting anomalies, and visualizations can tap into this. In Figure 1.4, we see a grid showing results of various elections; the voter turnout is arranged from left to right, the level of support for the winning candidate from bottom to top, and the color ranges from blue (this combination of turnout and winner support did not happen) through to dark red (this happened at many polling stations).

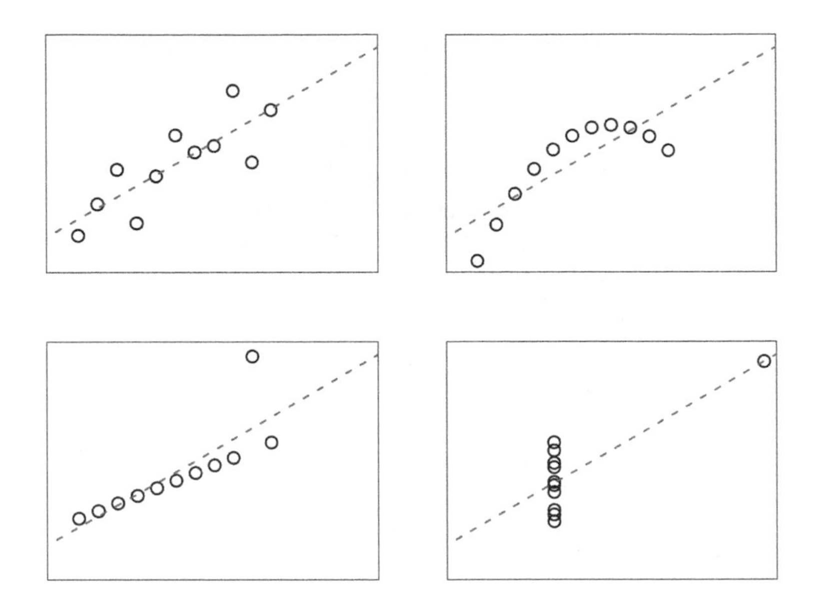

Figure 1.3 Anscombe's quartet: each plot shows a dataset with exactly the same statistics, masking great differences.

Elections in Russia and Uganda contrast with the others because of an isolated dot in the top right corner, where many polling stations had every registered voter turn up and every single one of them vote for the winner. The authors suggest this is evidence of fraudulent results because there is a gap between these stations and the rest of the stations. If we had to examine pages and pages of tabulated vote counts, we would have almost no chance of ever spotting this, but visualization makes it jump off the page.

1.3 TO WIN OVER THE AUDIENCE

Writing in 2012, statisticians Andrew Gelman and Antony Unwin suggested that there are at least six reasons why someone would create a visualization of data:

1. to give an overview,
2. to show the scale and complexity of the data,
3. to allow exploration of the data,

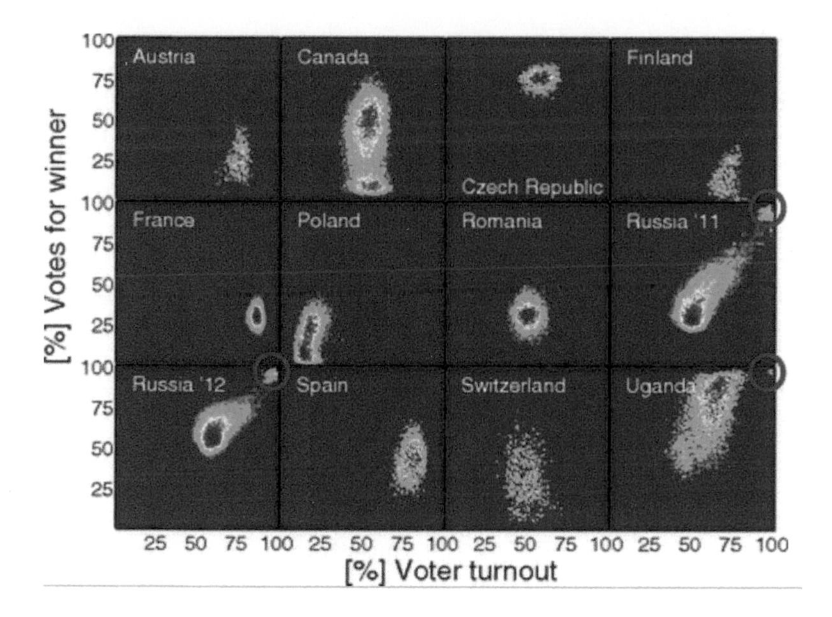

Figure 1.4 Voter turnout and winner support. Possible irregularities are circled in red. From "Statistical detection of systematic election irregularities" by Peter Klimek, Yuri Yegorov, Rudolf Hanel, and Stefan Thurner. Reproduced under open access publication license from the *Proceedings of the National Academy of Sciences.*

4. to communicate findings,
5. to tell a story, and
6. to attract attention and stimulate interest.

Reasons 1–3 are about showing data without interpretation, while 4–6 are about communicating a message. If you received some statistical training in the past, you were probably advised not to do numbers 5 or 6. Perhaps your teachers were themselves taught that data analysis must be scientific and objective, but you can be both these things while also being interesting and engaging your audience. Telling a story and catching the reader's eye is nothing to be ashamed of.

These six reasons would suggest multiple visualizations from the same dataset. There isn't one that's better than others, it's just that they have different goals. If different readers want to find out different levels of detail, then they will want to pick one from among these visualizations, and that means that one visualization is sometimes not enough. A top-level overview will work for some readers, while in-depth exploration is needed by others.

If you have findings to communicate, or recommendations to make on the basis of these data, you may need to tell a story. And if you fear that nobody will look at your work, you may have to catch their attention first. Layering the information from simple to complex allows users to drill down just as far as they want. Later in this book, we will look at interactive visualizations and how they potentially meet these different needs for all the readers in one place.

The term "infographics" is used for eye-catching diagrams which get a simple message across. They are very popular in advertising and can convey an impression of scientific, reliable information, but they are not the same thing as data visualization. An infographic will typically only convey a few numbers, and not use visual presentations to allow the reader to make comparisons of their own. Once you read Chapter 2, the difference should be clear. Figure 1.5 is a well-designed infographic: it is clearly laid out with just enough imagery to support the messages without getting in the way, it uses a limited range of colors that fit well together, and it conveys quite a lot of information in one place. But it is not a data visualization because it does not represent the numbers in any way except writing them out in text.

Throughout this book, there are many different formats for visualizations, from familiar ones such as bar charts to more exotic options such as hexagonal bin maps. When you have a dataset and a message to communicate, some formats will be better suited than others. A visualization that is confusing or hard to read can often be improved by using a different format. Sometimes the choice is obvious and sometimes there is no ideal format and the pros and cons have to be weighed against one another.

There are many possibilities, but here are some questions you can run through to help narrow down the candidates:

1. How many different observations and variables do I need to show?

2. Am I going to show the individual data, some aggregate of them, or some statistics (we'll look at this in Chapter 5)?

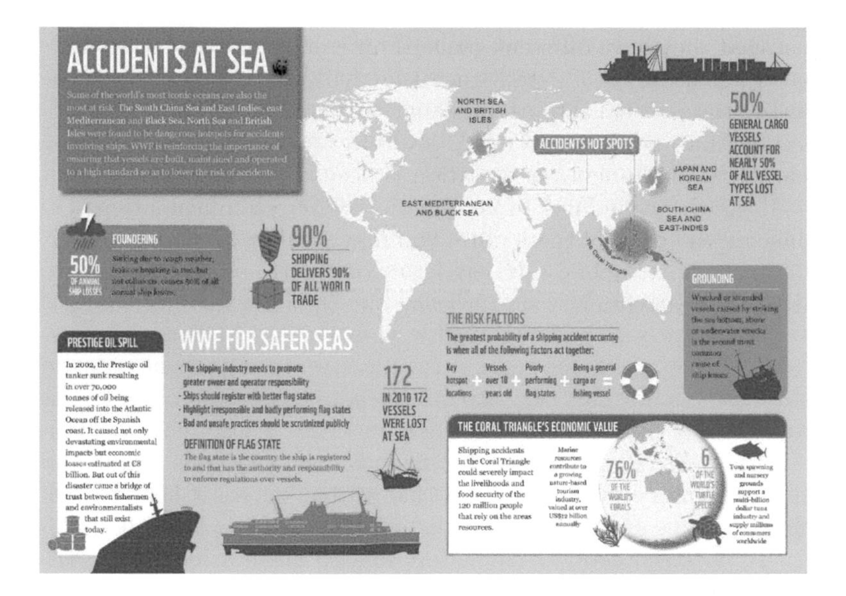

Figure 1.5 "Accidents at Sea," an example of an infographic. Reproduced with permission of World Wildlife Fund UK.

3. Are the variables nominal, ordinal or scale (these are dealt with in chapters 3 & 4)?
4. Do I want readers to focus on some comparison or change in the data (Chapter 6)?
5. What formats are my readers likely to be familiar with?

It is worth spending time on this and producing sketches with lots of alternative formats, because your goal is communication.

There is often no one "best" visualization, because it depends on context, what your audience already knows, how numerate or scientifically trained they are, what formats and conventions are regarded as standard in the particular field you're working in, the medium you can use, and so on. It's also partly scientific and partly artistic, so you get to express your own design style in it, which is what makes it so fascinating. Two people could take the same data and arrive at very different visualizations, both valid and useful (search online for "Tableau Makeover Mondays" for examples of this).

Imagine you work for an online retailer and you have asked customers to rate their experience with browsing your website, registering and paying, the delivery and the quality of the goods. Each of those gets rated from 1 (very bad) to 5 (very good). This is going to give a collection of percentages – and we'll go into them in detail in Chapter 4 – but that alone doesn't tell you what visualization to make. Maybe you want to just describe what the responses were, maybe you want to show what the company's strong and weak points are (people love our website but hate the delivery: time to switch delivery providers), maybe you want to compare changes from when you did the same survey last year, maybe you want to show how the ratings vary around the country, or maybe you want to show how people who didn't like the delivery also tended to dislike the website but love the goods once they got them. There's no simple answer in visualization. That's what makes it harder work than it seems at first, but it's also what makes it a lot of fun.

To choose a good visualization, and to make it look clear, understandable and attractive, you need to know a little about how we see and process vision – the brain power described above – and basic design principles, and we will discuss these in Chapter 7. If you are aiming to communicate a message, you also need to have a very clear idea of what that message is. It is easy for a visualization to become confusing because too many messages are being squeezed in. Here are some more questions to ask as you begin work on a visualization:

1. What is the message?
2. What parts of the data are evidence for it?
3. What other parts need to be shown for contrast/context?
4. Do I/we know how to do this, or can we learn it / adapt someone else's work, or do we need to hire in?

There are some websites and publications that show a typology of visualizations, listing the many common formats and what they are useful for. Some of these are really nicely designed, but I don't recommend them if you are learning data visualization or trying to choose a look for a particular problem. I think that each problem is best served by thinking about what you need to show and how best to encode and format it. There are several instances in this book where I show something that you won't find in any typology of standard formats.

1.4 WORKING ON DATA VISUALIZATION

If you are the boss in an organization, you have data, and need to produce a data visualization, consider first whether you have the necessary skills in your team. On the design side, you may be able to access people in-house like web developers and graphic designers who would enjoy a different challenge for a while, and benefit from having a new skill on their résumé.

For statistics, you may already have people who work with data analysis software, and even if they do not typically produce graphical outputs, they might also enjoy learning something new. In either case, ask them whether they enjoy teaching themselves new skills quickly, and if that is so, your own team can probably get the job done. However, there is always a risk of it not looking very professional or taking longer than you hoped for, so if you have a budget for it, you can hire in someone to do the work on a consultancy basis.

There are many studios and agencies that would consider the work, provided the data are fully cleaned, analyzed and ready to go. Here are a few pointers to guide that decision:

- The individuals who take on the job will come from either a statistical or a design background; think about which of these you need. If your data analysis is complete and you want a few slick visualizations to achieve the goal of catching the audience's eye, then it is design you need. If you already have a very well-specified communications policy with branding thoroughly defined, but the data need to be processed and represented fairly, then it is statistics (and possibly programming and database skills as well). If you need both, you will need to hire an organization that is large enough to provide both. Trying to act as a middleman between two contractors from very different backgrounds will not be easy. On the other hand, if you need a very fast turn-around, hiring a single self-employed consultant is probably the best option.

- The people you hire can only do what you tell them you need, so clearly and comprehensively specifying the requirements is essential. Never assume that they have understood something. There is no harm in giving a lot of detail, but not enough detail will risk delays and expense.

- Expect to be involved in user-testing, to look at early drafts and provide feedback. It is essential to test and refine visualizations because there is no one right way to do them, but they must match the requirements and knowledge of the audience. If you are hiring a small studio or an individual, you may have to organize a group of your staff, customers or other contacts to act as a focus group for them, so that they can get relevant feedback on drafts.

- Do not underestimate the tasks of design or web development. They may look simple, but the detail is hidden from view. It is probably not possible to squeeze in a good idea later that wasn't specified up front, without incurring delays. Also, if you really cherish one idea that you had, and the designer keeps telling you it won't work, accept that they know more about this than you do, and let it go.

If, on the other hand, you are thinking of taking on work making visualizations for other people, this advice applies:

- Find out what the data are like before you start. Will analysis be required; can you do that or can you get someone to do it as a sub-contractor? Is everything labeled, defined, coded consistently and available to you on day one? Ask to see the data before giving a quote, offering to sign confidentiality agreements if necessary. It shows that you are serious about helping the organization.

- Get the timescale, budget, specifications and any confidentiality requirements agreed in writing first. Ask for copies of any communications policy or branding guidelines from the organization hiring you, and make sure they stick to this.

- Propose a program of user-testing and feedback from the outset, and get agreement to meeting times for that. It is all right to identify problems – that is part of the design process – and you can use these meetings to agree on how to deal with them.

- Throughout, remember that your relationship to your clients is a professional one: they want you do a great job and to tell them what would be a good or bad idea. They might push for a bad idea until convinced otherwise, but it is your job to give them straight-talking advice. Do not be tempted to cut corners or allow mission creep so that you can secure the job; be prepared to walk away from a deal that will cause you headaches. Similarly, clients who are not trained in statistics may ask for you to "massage" the figures a little to favor their point of view. You should always help them to understand why this is a bad idea: quite aside from misleading and maybe harming people, the damage to both your reputations is not worth it.

1.5 A TOOLBOX

Because these things change rapidly, I don't talk about software in this book. The code to make the figures is on the accompanying website. But I will say this at the outset: if you want to become a versatile visualization maker, I recommend building familiarity with a box of tools like this:

- Programmable statistical software like Stata or a data-focused programming language like R or Python,
- quick visualization software like Tableau,
- SVG editing in the text editor as well as a graphical interface like Inkscape or Illustrator,
- sketching by hand,
- relevant JavaScript libraries – currently, D3 and Leaflet would be the choice,
- an online mapping tool, like Mapbox,
- a versatile program for big data and fast data, like Spark.

I have not mentioned spreadsheets here. This is not because I am a snob, but because it is far too easy to make mistakes with them, and too hard to track those mistakes down. I recommend you avoid them.

1.6 BE PREPARED TO SKETCH AND DISCARD

Because you need to choose from lots of competing designs and formats, you will need to sketch some and discuss them. Often, the quickest and most flexible way to do this is by hand – not with any software. Your sketches will mostly be thrown away and one or

Figure 1.6 Step one in sketching an idea for visualizing data on the distance traveled by commuters in the city of Atlanta, Georgia, United States: I used tracing paper to get minimal information from a printed map. I chose the city boundaries, shading for built-up areas, blue for reservoirs and lakes, and orange for major roads. This should be enough information for local people to recognize (see Chapter 13 for more on maps), and I could use the orange to signify travel again later. We revisit this dataset in Chapter 3.

two will emerge victorious to inform the serious work that follows. They do not have to be beautiful – they have to be clear about what goes where.

Quickly do lots of them and throw all but the best ones away. It is worth thinking a little about color at this stage too, which is the perfect excuse for having a large box of crayons on your desk. Graph paper (which comes printed with an accurate grid) is not important for this purpose, but it is worth having some acetate (overhead projector) sheets and a set of colored glass marker pens with fine points. You can use these to overlay extra information on your sketch and then change your mind without having to redraw everything. The very first ideas can also take shape on a whiteboard using these pens. Tracing paper is also handy for getting map outlines or copying existing visualizations with designs that you like.

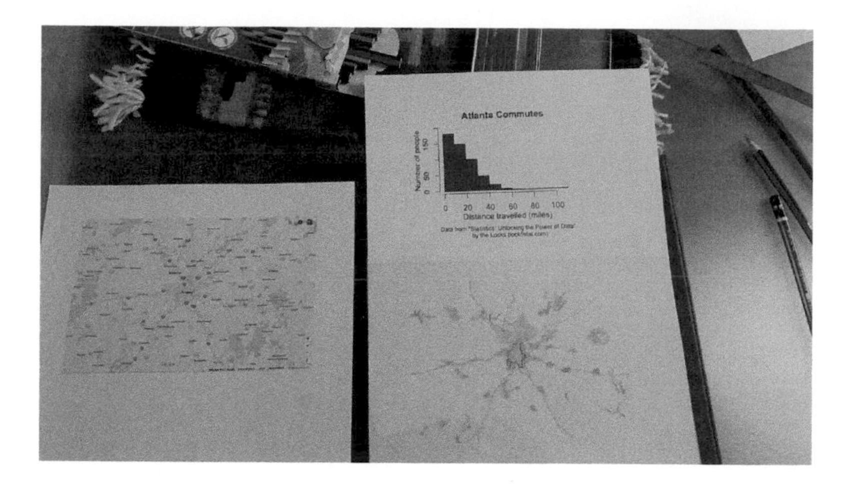

Figure 1.7 Step two: I attached the tracing paper to a blank sheet of paper, and an acetate transparency sheet on top. Map data and image copyright Google.

The worked example below takes a simple dataset on commuters in Atlanta, Georgia, United States, and explores one way of making it more engaging and meaningful. The intended audience is local commuters themselves. After choosing a few designs and creating them with appropriate software, the next step would be user testing, to overcome any assumptions and biases that we might have, and to choose a final format for publication. I cannot stress the value of user testing enough!

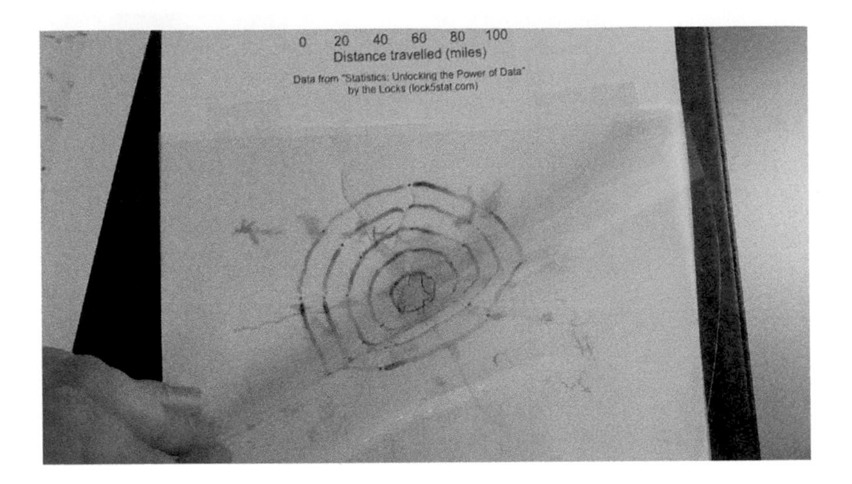

Figure 1.8 Step three: I drew rings on the transparency at ten-mile intervals around a central point. This relates the division of data directly to the real world. Map data and image copyright Google.

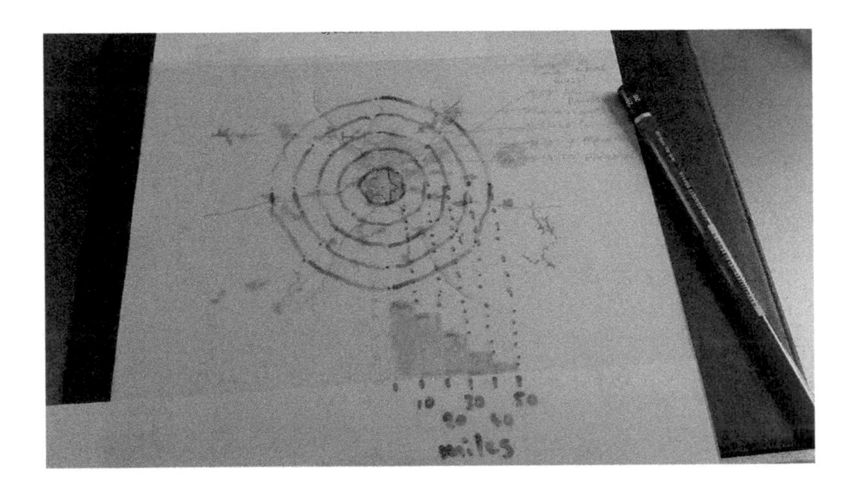

Figure 1.9 Step four: I added annotations to the sheet underneath. I decided that I didn't want to clutter the map with names of towns, but it would be helpful to local readers to see some of the main places in each ring named. Map data and image copyright Google.

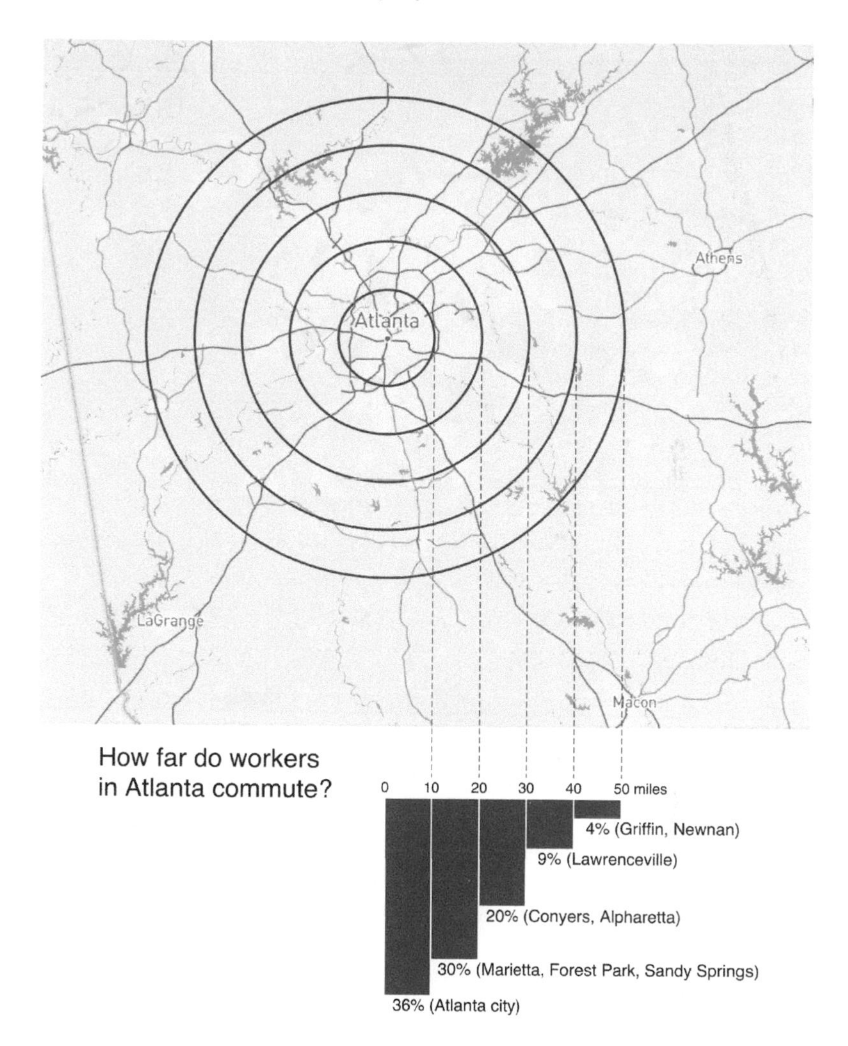

Figure 1.10 A first digital draft of a contextualized bar chart of commuting distances in Atlanta, Georgia. I decided to leave extra space on the map outside the 50-mile radius, so that the rings and the bar chart were visually spaced apart, linked just by the dashed vertical lines. Empty space can be valuable. This is ready to be given to some commuters for feedback, to check that it is clear and understood. Data from *Statistics: Unlocking the Power of Data* by the Locks, map image copyright Mapbox, map data copyright OpenStreetMap contributors.

CHAPTER 2

Translating numbers to images

DATA VISUALIZATION pioneer William Cleveland wrote that visualizations translate numbers into images on the page or screen, so that the reader can translate them back again in their brain. Both of those translations have to work, and there are many ways in which we can take a number and make it into some aspect of an image, so the choice has to be made carefully. Also, the reader should be able to so without craning their neck, scratching their head and generally struggling over the image, or they may well just give up and move on.

Only a few seconds of confusion will dissuade most of your audience, because of short attention spans, other demands on their time, and math-phobia. But look at it from their perspective: if you have something important to show them, get on with it, don't make them work for it. They have to be able to translate it back to numbers, and do so quickly. Needless to say, they should also arrive at the correct number and not be misled by the visualization.

2.1 LEAVES ON THE LINE: AN EXAMPLE OF VISUAL ENCODING

This chapter explores an important concept that runs through data visualization: when we do that translation into images, we have to connect up values in our data with parameters of the image. This is known as visual encoding of data. The best way to get to grips with this abstract-sounding notion is to jump right in and try out a simple example.

Let's explore whether delays on trains in South East England are worst in the autumn, when the rails are slimy with leaves falling from trees alongside the track, and the trains have to slow right down to be sure they can stop safely. This "leaves on the line" problem has attained legendary status in England, but is it really true? I downloaded publicly available performance data from dataportal.orr.gov.uk and tidied it up, but you can download it directly from the website for this book at robertgrantstats.co.uk/dataviz-book.

The variables we have to consider are time (the data are available for 4-week periods, and there are about 13 of these in each year) and the percentage of train journeys recorded as being cancelled or "significantly delayed." We can also divide time up into years and 4-week periods within years, or we could identify autumn versus the rest of the year. The data covers twenty years from April 1997 to March 2017.

A scatter plot is one obvious choice, with time encoded to the horizontal location, and delays encoded to the vertical location. As shown in Figure 2.1, we have encoded our two variables (time and

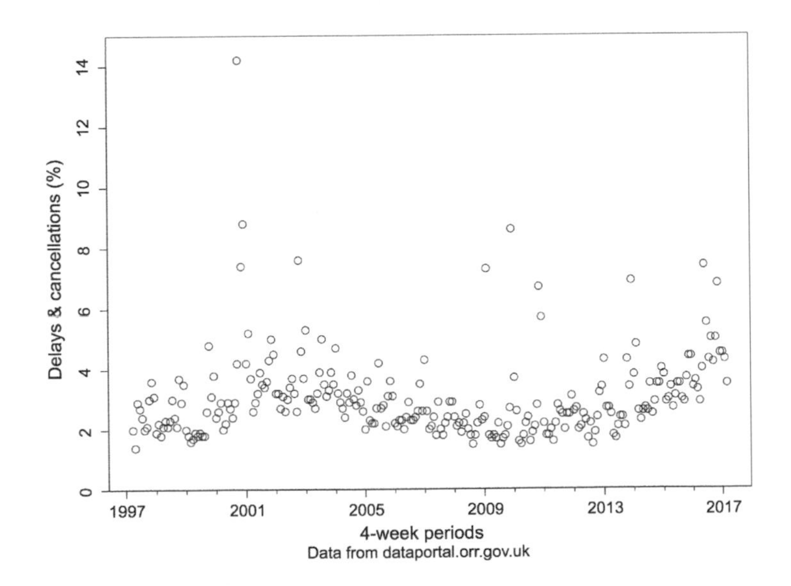

Figure 2.1 Train delays: a first scatter plot

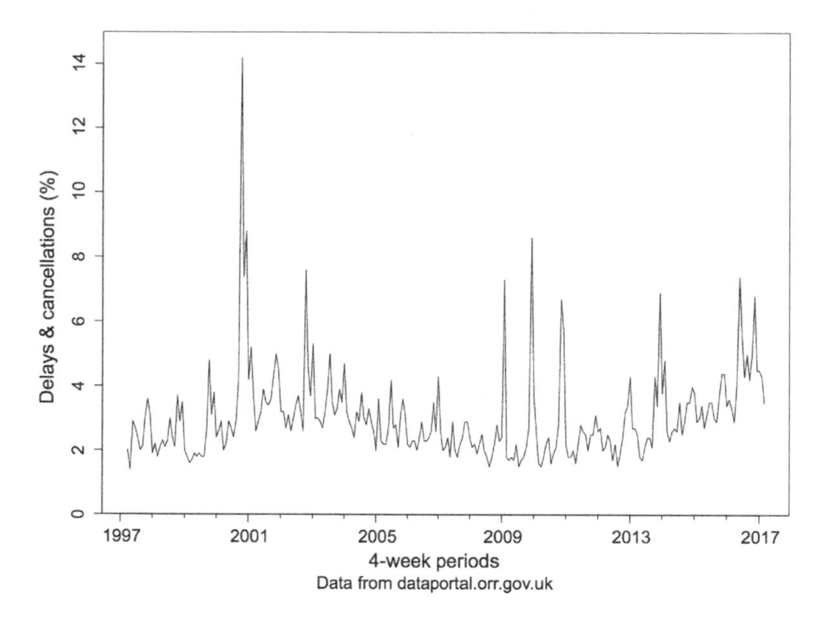

Figure 2.2 Train delays: a line chart

delays/cancellations) to two image parameters, and then drawn a marker for each of the observations at the resulting location. The % delays and cancellations are values close to zero, so the vertical axis has been extended to 0%. This helps to convey the values of the markers as their height in the chart. However, this has not been done for the horizontal axis because there is no meaningful zero value in this context.

Encoding time to the horizontal location makes intuitive sense to many readers because we read from left to right (in Latin alphabet languages anyway). We can easily spot the general trend: getting worse, then better, then worse again. We can also see the especially bad periods as high points on the chart, but it is hard to compare autumn in different years because they are scattered across the plot.

An alternative format with the same encoding is the line chart (Figure 2.2). Although the same variables are encoded to the same visual parameters, we connect the data with a line rather than drawing individual markers.

This has an effect on how it is perceived; you might immediately notice that the really high points are more obvious, and the long-term trend less so. The high points require the line to go all the way up to a spike and all the way back down again, thus putting a lot of "ink" on the page (or screen), which draws the reader's eye in a way that doesn't happen for the scatter plot.

Statistically speaking, the delay/cancellation percentage (%) data are positively skewed: most of the values are around the same value, a few are much higher, but there are none that are much lower. This sort of distribution can inform your choice of encoding and format. We will look at distributions in more detail in Chapter 3.

How to read a line chart

Line charts almost always trace lines from left to right, with the height of the line indicating the value of principal interest. They often, but not always, have time on the horizontal axis so can be read as a sequence of events. High points, low points, periods of (in)stability and lines crossing other lines are all of potential interest to the reader and relatively easy to pick out. When there are too many lines, and each one moves up and down a lot, it can quickly become too cluttered to be read reliably (often referred to as spaghetti by dataviz people). One solution to this is to use "small multiples," which we will revisit in Chapter 12.

Sometimes, a line chart uses one vertical axis on the left and another on the right, and their values relate to different lines. This is generally a bad idea because it encodes two variables to the same parameter – vertical location – leaving it to the reader to distinguish between them. Always try to make things as easy as possible for your reader.

The story behind these visualizations is all about comparing autumn to other times of the year, so we need the reader to be able to identify autumn. One approach would be to highlight it, which I have done in two ways in Figure 2.3. On the left, one period each year (mid-November to mid-December) is picked out with a colored dot, and on the right, colored shading is used to identify mid-September through to mid-December.

The dots have the advantage of showing you exactly where the autumn data lie, but it starts to get too visually busy to read easily. Shading is less intrusive on the data markers, but the candy-stripe effect creates its own kind of visual overload.

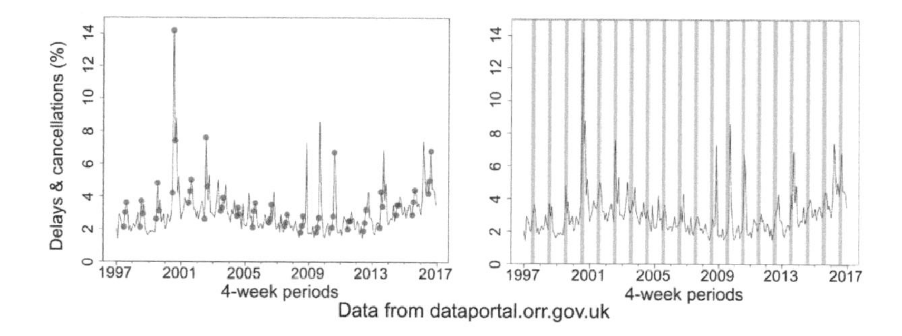

Figure 2.3 Train delays: a line chart with highlighting. The points identifying autumn in each year might work, but the candy-stripes are distracting and overload the reader.

The color, by the way, was picked from a red maple leaf in real life using a smartphone app. We'll think about aesthetic choices like that in Chapter 16.

At this stage, some new messages start to emerge, thanks to our choices of encoding and format: although the autumn was indeed the worst time for rail travel in most years up to 2002, from 2003 onwards the worst period has tended to be later, in the depths of winter. Perhaps leaves have been defeated but ice remains a problem. Without visualization, could you have spotted this from the raw data?

In Figure 2.3, I effectively created a new variable, which just identifies autumn. This is just a variant on the time variable that we already had. We could take another approach to help readers compare autumn to the rest of the year, again making a variant on the time variable by splitting it into the year and the 4-week period. Then, we can encode the 4-week period to the horizontal position (Figure 2.4).

Figure 2.4 has definitely not helped! In the scatter plot, we can no longer tell which year a particular point belongs to, and in the line chart, even though there is one line per year, they intermingle so much that it is impossible to trace any line across the chart. This is because we now have three variables that have to be encoded: year, 4-week period, and % delayed, and we are neglecting the year.

We have already used the horizontal and vertical positions, so we need another chart parameter to indicate which year each marker or line belongs to. One option is to use different colors, but

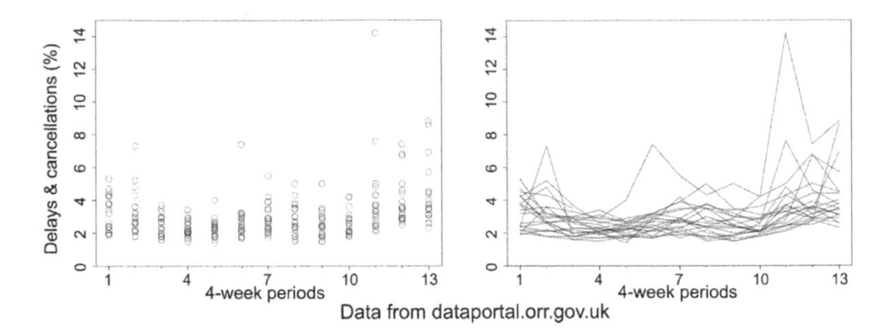

Figure 2.4 Train delays: an unhelpful scatter plot and line chart for periods within each year.

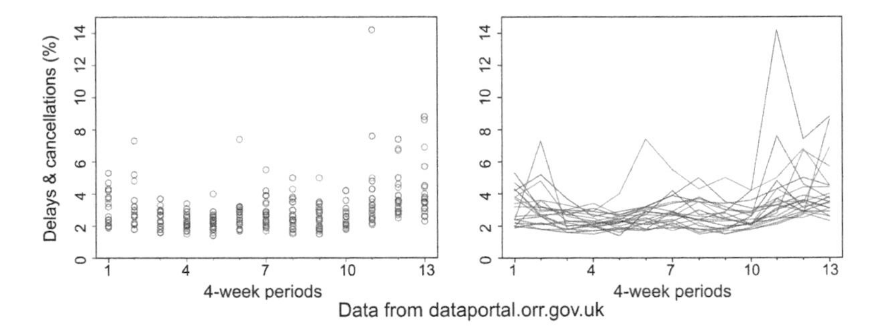

Figure 2.5 Train delays: scatter plot and line chart with old data in black, ranging to newest data in orange

we must allocate colors that have some sense of going in a particular order, not just a rainbow effect, otherwise the order of the years will be lost (Figure 2.5).

This helps a little in the line chart (compare the highest lines at period 10 with the same lines in Figure 2.4), but not at all in the scatter plot. If we had only a few years, the colors might be so distinct that we could tell each one apart. But the fundamental problem here is that the years are all quite similar and so their lines lie on top of one another: a statistician might say that the variance *within* years is larger than the variance *between* years. You have to take this into account if you want the message to be clear and not swamped by irrelevant noise.

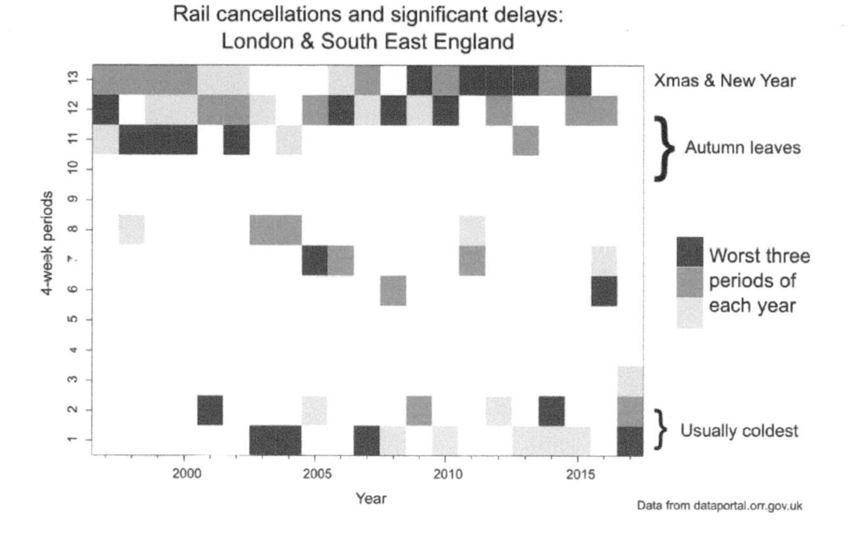

Figure 2.6 Train delays: year-period grid, with the worst 4-week periods in each year colored

There are some tricks we can employ to help us out of this difficult spot that fall under formats rather than encoding, which I will return to later in the book: we can smooth out the lines (Chapter 5), show multiple small charts, one for each year (Chapter 12), break the data down with statistical time series analysis (Chapter 9), or present it digitally as an animation or an interactive web page (Chapter 14).

I want to close this example by breaking out of the horizontal-vertical encoding of delays and time that we have used throughout. Let's think about the message again: we want to explore the "leaves on the line" effect, so we need to focus on the time of year when delays and cancellations are really bad. Perhaps the actual delay/cancellation % does not matter as much as just identifying the worst periods in each year. In Figure 2.6, the data are shown as a colored grid, with the year encoded to the horizontal position, the 4-week period to the vertical position, and the delay/cancellation % to color. To avoid the color overload we saw before, only the three worst periods in each year are colored.

Grids of color like this (or Figure 1.4) are often called heatmaps. This has the advantage that the story we want to convey will be front and center. If there is a consistent autumn leaves effect, it will present itself as an obvious horizontal stripe of color. The actual % in each of the worst periods is ignored, and they are simply colored according to whether they are the worst, 2nd worst, or 3rd worst for each year, and everything else is left white. This deliberately throws away a lot of the information in the data in order to achieve clarity. That is not a decision to be taken lightly, but in Chapter 16 I put this chart into a poster together with more detailed information to back it up.

At the end of this extended example, you should understand the difference between choosing formats and encoding data to visual parameters. With the final chart of train delays, we can see that problems have shifted from autumn to the coldest months of the winter – December to February.

You should also realize that the choice of encoding and format is tied up with what the message is that you want to communicate; there is no universal right answer for one set of data.

2.2 CHOOSING VISUAL PARAMETERS

There are many parameters we can encode to. Research into human visual perception has shown that some are not so easy for readers to translate back into numbers. Areas, for example, are usually perceived as being more similar to each other than they really are. We'll explore this further in Chapter 7. In Table 2.1, parameters known to have problems like this are identified with asterisks.

Using volume to depict a quantitative variable, for example, is not ideal because it is very hard to judge the precise relative size of each "object" as depicted on the page or screen. Volumes could be used for ordinal data, or a different parameter such as length would be better suited. There are, however, exceptions to any rule in dataviz, and for an example of making good use of volumes to convey the scale of the data and attract attention, consider Figure 16.5.

In the Preface to this book, I defined variables as categorical (with ordinal as a special case) and quantitative. Some of these parameters are suited only to having non-ordinal categories encoded to them, while others will work for ordinal values but not the other two.

Being measured against a common scale is important. It is quite easy to judge the relative lengths of three lines which start at the same point, like in the table above, but much harder if they are not adjacent, in other words, they differ in horizontal location too, let

TABLE 2.1 Visual parameters

Parameter	Ideal data type	
Position	Quantitative	
Length	Quantitative	
Angle	Quantitative*	
Area	Quantitative*	
Volume	Quantitative*	
Color hue	Nominal or ordinal*	
Color saturation	Ordinal	
Marker shape	Nominal*	
Features	Quantitative*	
Line width	Ordinal*	

alone when they differ in other parameters such as angle or width (Figure 2.7). This is one reason why pie charts are hard to translate back to values with an acceptable level of accuracy. There are other circular designs such as "radial histograms" and "sunburst diagrams" that suffer the same problem. They may be eye catching and supply the sixth of Gelman and Unwin's objectives, but fall short for serious quantitative communication, and are not as novel and exciting as some of the closing examples in Chapter 17.

Yes, it is possible to see that each case in Figure 2.7 is in the ratio 1:3:2 after a few seconds of concentration at most, but we know that readers of visualization might not concentrate at all

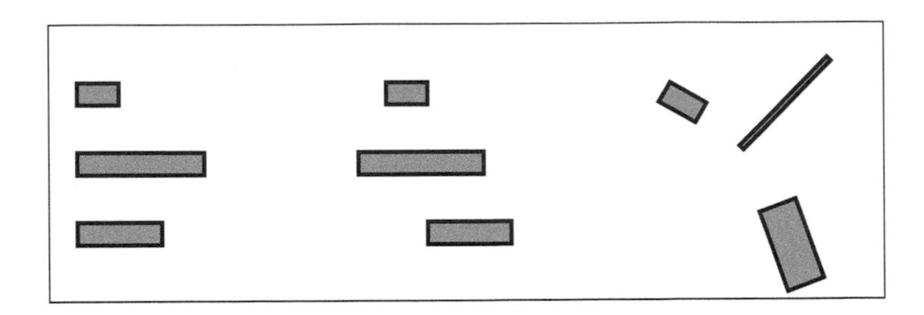

Figure 2.7 To let the reader translate a parameter back into numbers, keep everything else constant. The lengths of the three bars on the left can be compared, but less so for the three in the center, where the baseline position has been shifted, or on the right where the width and angle changes too.

but rather draw conclusions based on first impressions, so it is important to eliminate any chance for misunderstanding like this.

The lesson of this is that, if comparing some values is important, those values should be prominent and adjacent. It can be tempting, with many variables, to encode each to some parameter and produce a very detailed, complex visualization. We can make a scatter plot where each marker has not only a horizontal and vertical location but also a particular shape, color and size – five variables in one chart – but we should not be too surprised to find that many readers would be discouraged by this. When there are important messages involving specific variables and specific values, it is best to apply the layering principle from Chapter 1: single them out visually in a headline chart and then present the rest of the detail elsewhere.

How to read a bubble chart

A bubble chart, like Figure 13.5, is just a scatter plot with a third, quantitative variable encoded to the size of the marker. This is particularly useful because sometimes we want to emphasize that some of the observations in our data contain more information than others. Imagine plotting data on countries – it would make sense to give more emphasis to high-population countries in a visualization. There are two major problems with bubble plots, though. Firstly, the bubbles sometimes overlap and obscure one another. You might find that drawing just the outline of the bubble and not filling it with color solves this, or you can make them semi-transparent. Secondly, we know that areas are not accurately perceived, so if the third variable matters a lot, it is probably a good idea to choose a different format or encoding.

Not quite every combination is possible, but there are many that will lead to unusual and possibly innovative visualizations. However, encodings that are already familiar to your readers allow them to skip those precious few seconds where they have to work out how to read the visualization, and lets them move straight to understanding the data. Innovation is always worth considering, but the primary goal of most data visualization is to communicate. We could move away from the horizontal-vertical alignment completely and do something more meaningful by tapping into the circular nature of the seasons. The 4-week periods could be arranged in a circle or a spiral instead of a straight line, and we'll look at formats like that in Chapter 9 on time series.

2.3 UNDERSTANDING COLOR

Of all the image parameters we can encode to, it's worth learning about how we can quantify color. There are several systems for specifying an exact color, but this book will mostly use the RGB (red, green, blue) system. Nearly seventeen million colors can be described by giving a number from 0 to 255 to the amount of red, green and blue that get mixed together. So, the darkest rectangles in Figure 2.6 have RGB values of 173, 56, 41. The red component (173) is the largest number, with relatively little green (56) and blue (41). Pure black is 0, 0, 0 and pure white 255, 255, 255.

This means that any color can be described by three numbers. We'll return to RGB when considering visual perception and the brain in Chapter 7. Another notation system called HSV describes colors in terms of hue (basic colors from the spectrum), saturation and value (added lightness or darkness) – terms which appear in Table 2.1.

Whatever we use, we need to accept that the human brain is not good at judging distances between colors in terms of mixing together red, green and blue, so encoding three variables from your data to these primary colors is a bad idea, tempting though it may seem.

Likewise, if color really is part of our dataset, then it is probably a good idea to simply draw the color in question (encode it as itself) in the visualization (although we'll explore this further in the example below), and not to three other parameters. However, a more sparing use of color can work well, not encoding to the whole spectrum but to shades of one color, like the saturation option in Table 2.1, or the right-hand part of the hue option.

Despite our inability to see "distances" between colors, we can still tell what is lighter or darker at a glance (ordinal information), and this may be all that is needed sometimes, as in Figure 2.6.

2.4 THE LIMITATIONS OF AREAS

One very common problem in data visualization is that encoding numerical variables to area is incredibly popular, but readers can't translate it back very well. Bubble charts, which we've already encountered in this chapter, suffer from this. The reader can probably spot the biggest and smallest markers, but will find it hard to judge sizes of others that are not adjacent. The same problem affects pictograms where a variable is encoded to the relative size of some icon. The area is not proportionate to the height of the icon or bubble, but that is how we generally see it.

Wordclouds take this problem deeper into inadequacy, with the human reader hardly capable of judging the relative areas of words of different lengths, with different sized characters, in different colors and different rotations. A better idea, if you want to use icons, is to line up a number of them (Figure 2.8). This is also called a pictogram, and sometimes an Isotype chart, named after its early proponents, the Isotype Institute. Because here, the variable is encoded to length, it is much easier to see things clearly.

There's another special kind of pictogram that is not often seen, perhaps because it takes a lot of effort to produce, where the icons are real objects in a photograph. Figure 2.10 is a photomontage, combining separate photographs of every plane taking off from Los Angeles International Airport's South Complex in a day, while Figure 2.11 was posed in real life with the city's permission.

Because they rely on areas, it has been shown that readers regularly misjudge relative sizes of the slices in pie charts and donut charts. A slice at the top or bottom of the pie (or donut) will appear smaller than the same sized slice at the left or right. Also, people

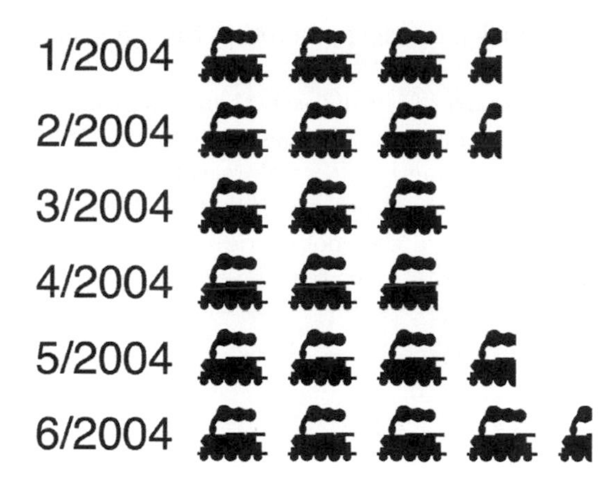

Each icon represents 1% of trains cancelled or delayed.

Figure 2.8 A pictogram in the Isotype tradition. Train icon vectorized by Wikimedia user "Sgt bilko," drawn by Wikimedia user "Eschweiler," CC BY-SA 3.0, https://commons.wikimedia.org/w/index.php?curid=1158450

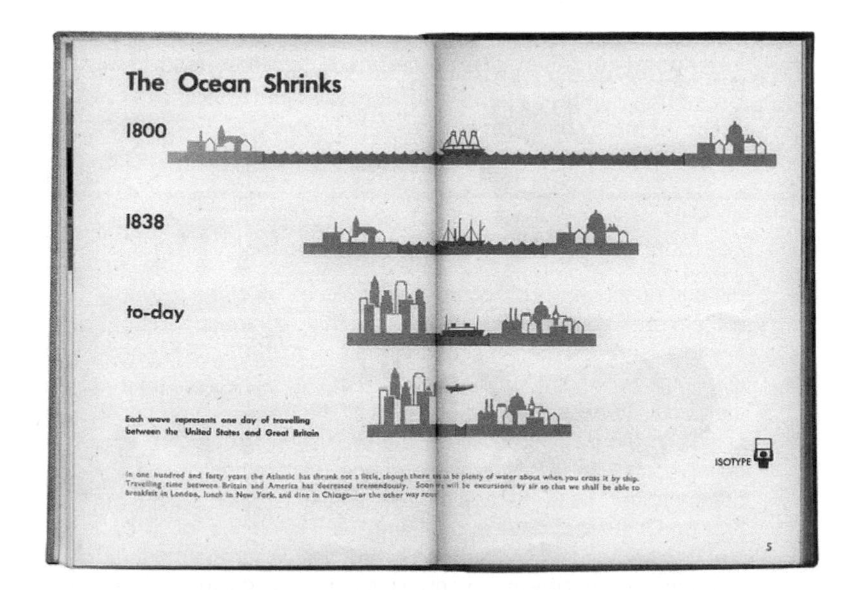

Figure 2.9 "The ocean shrinks," a visualization by Otto and Marie Neurath of the Isotype Institute. Public domain.

Figure 2.10 "Wake Turbulence" by photographer Mike Kelley visualizes every plane to take off from the LAX South Complex in a day.

typically underestimate angles less than 90 degrees and overestimate those which are larger.

Different colors and 3-D effects can trick the reader too — we'll explore these further in Chapter 7 on visual perception and the brain. Although angles are proportionate to areas (if the radius stays the same), the problem of different orientations affects the pie and the donut (see Figure 2.7).

How to read a pie chart or donut chart

Pie and donut charts divide up a circle like slices of the proverbial pie, and so are well suited to data that represent parts of a whole, perhaps percentages that add up to 100%. The only difference between them is stylistic, with the center omitted in the donut. In both cases, the angles of the slices are proportionate to the data, so a 90-degree slice should represent an observation with twice the value of one represented by a 45-degree slice. The 90-degree slice will also have twice the area, provided that slices have the same radius. The number of slices that can be accurately displayed is probably fewer than ten. A pie chart with dozens of thin slices is of very little use for data visualization, and new problems are introduced, such as having

Figure 2.11 Road congestion for different forms of transport. A physical visualization by the International Sustainability Institute. Reproduced with permission.

to use several similar colors or squeeze in labels for the slices. If you really do want to convey the message that there are many categories in your data, just say so in text.

If you want to do accurate dataviz that helps your readers and doesn't attract criticism, you should try to avoid pies and donuts where possible. Also, they are widely ridiculed in the dataviz world, which might matter to you if you want to build a reputation for good practice. Two good alternatives are the familiar bar chart and the waffle plot. The bar chart simply encodes a quantitative variable to the lengths of rectangles. They can easily be compared because they share a common scale, as seen in Figure 2.7. If you want to convey the impression of how much each bar contributes to the whole, you can have all bars extend to the 100% mark and color them in to the proportion that represents each observation in your data.

The same information is conveyed by replacing the bar with a line, and emphasizing the extent to which it is filled by superimposed circles; this is called a dot plot, but that term is also used for a different chart called a dot plot, which you will encounter from Chapter 3 onwards.

The waffle plot is a special kind of pictogram, where little squares are lined up close together. It is similar to a bar chart in that we can look at the length of a row of squares, but if the values in our data differ greatly, then we could arrange the squares in two dimensions. This has the advantage of accommodating much larger numbers. We could show three categories in the ratio 1:3:2, like in Figure 2.12, but we could also show 1:3:2000 (Figure 2.13).

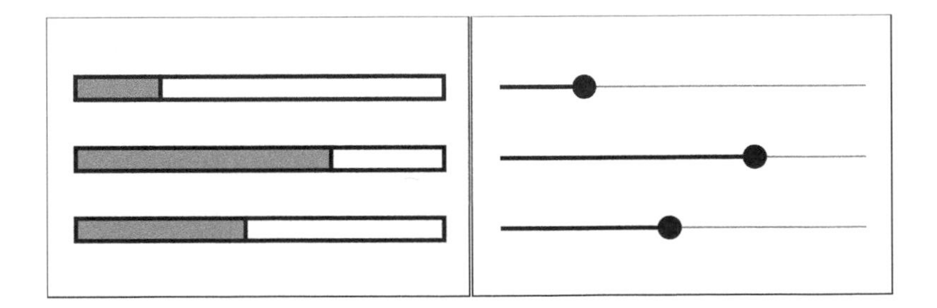

Figure 2.12 Proportions of the whole in a bar chart (left) and a dot plot (right)

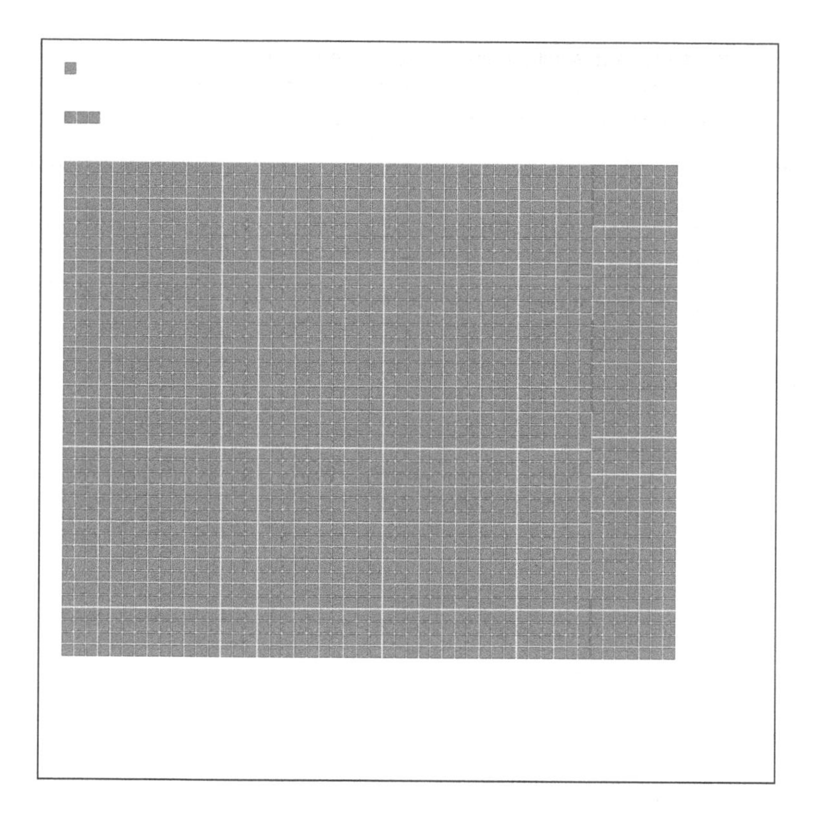

Figure 2.13 Very different values side by side in a waffle plot: 1, 3 and 2000

In a bar chart, this would make the first two values so small, they would be invisible.

How to read a bar chart or waffle plot

A bar chart is probably the simplest of all formats to read: the length of the bar is proportionate to the value that is encoded to it. A waffle can go over onto multiple rows and then the area is proportionate. It is the fact that one could potentially count the squares that makes this better than the pie or donut.

Finally, why are some formats called charts and others plots? There's no reason, it's just convention.

2.5 ANNOTATION

Our goal is for our visualizations to be clearly understood, and some annotation with text can help. Unfortunately, if we pile in too much text, with attendant arrows and highlighting, it will quickly make the overall impression too busy and confusing. Edward Tufte's advice to reduce the ink:data ratio is worth bearing in mind at all times – but that doesn't mean that the absolute minimum is always best.

A good general rule is to imagine the visualization being read apart from any accompanying documentation or context: will it be misunderstood? Labeling axes and saying where the data came from is vital, and many visualizations benefit from a title that tells the headline message. You don't have to be po-faced: notice how the Isotype image in Figure 2.9 is titled "The ocean shrinks," not "Travel times by different forms of transport." I'll return to annotation in Chapter 16.

2.6 USER TESTING

This chapter has introduced various visual encodings and formats. There will be plenty more in the chapters that follow. Whatever approach is taken, an effective data visualization is only guaranteed if it works for the intended readership. When you are looking at someone else's dataviz, consider what you like and don't like about it, and then how you might improve on it. This way you will learn quickly from others' mistakes.

If you find one confusing, don't assume that it is your fault: visualizations should be simple and intuitive. If you are making dataviz, or supervising someone who is, then you should definitely consider user-testing. Show drafts to some people who are like your intended audience and get some honest opinions from them. The user experience – ease of use, accessibility and enjoyment – matters a lot, so the whole package of the image, accompanying text, color schemes, annotations and terminology should work together to get the message across and allow them to explore the data visually. We will explore those aspects further in Chapter 16.

Although standard, familiar formats are powerful, there is an attraction to novelty and some readers can be drawn in because of curiosity about something new. Don't shy away from innovation, but be careful to user-test it.

II

Statistical building blocks

CHAPTER 3

Continuous and discrete numbers

THE BEST PLACE TO START learning about data visualization is with quantitative variables. By quantitative, I mean that they can take a range of numeric values that measure something in the real world. We can split them further into continuous variables and discrete variables. Continuous variables can take a very precise value to many decimal places, at least in theory. Discrete variables can only take particular values, usually 0, 1, 2, 3... (what mathematicians call natural numbers).

A good way to spot these variables is that if there are units of measurement (for example, acres burned in a forest fire, or hours of battery life for a new phone), it is definitely continuous or discrete. Unfortunately, the opposite is not true: there are also discrete and continuous variables with no units, such as the ratio of men to women among job applicants.

The other form of data is categorical, which indicates that each observation falls into one category. Those categories can sometimes have a natural order to them, but that doesn't make them discrete. Discrete variables have a true numeric meaning: you can subtract the population of Anytown from that of Othertown, and it means the number of extra people. That is not true of ordinal categories like Strongly Disagree, Disagree, Agree, and Strongly Agree, even if you record them as numbers like 1, 2, 3 and 4. We'll come back to them in Chapter 4.

3.1 ONE VARIABLE AT A TIME

Let's start with one variable containing continuous data: the distance that people commute to work in the city of Atlanta, Georgia, United States. Before we draw anything, even before looking at the data, there are some aspects we expect to find. The numbers have to be positive, and we would expect most of them to be under twenty miles, with a few much longer.

You don't have to know anything specific about Atlanta to realize this, but it's useful to stop and think about what you *expect* to see before you get going. When you create visualizations, if you find things not quite as you expected, it might suggest you did something wrong – which is very easy with data analysis software and spreadsheets.

It feels sensible to encode the commuting distance to the horizontal position, from zero miles on the left to the maximum commute distance on the right. Let's draw a dot for each person in this data set, and stack them up if they are within 2 miles of each other's distance. Figure 3.1 is sometimes called a strip chart, and

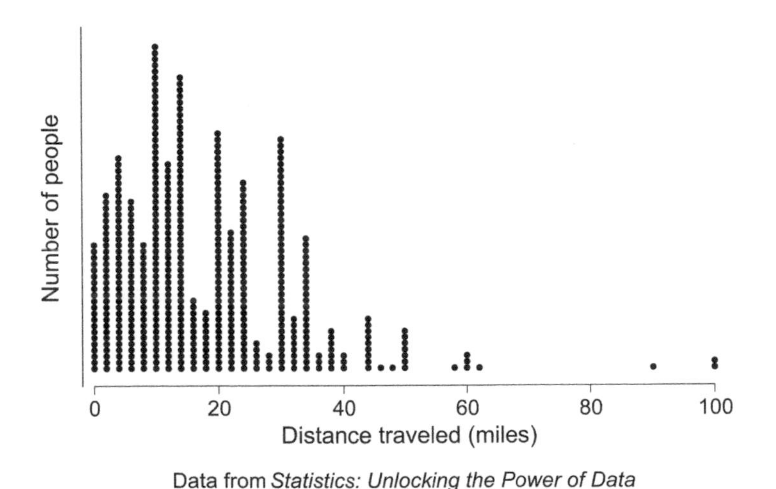

Figure 3.1 A strip chart or dot plot of one continuous variable.

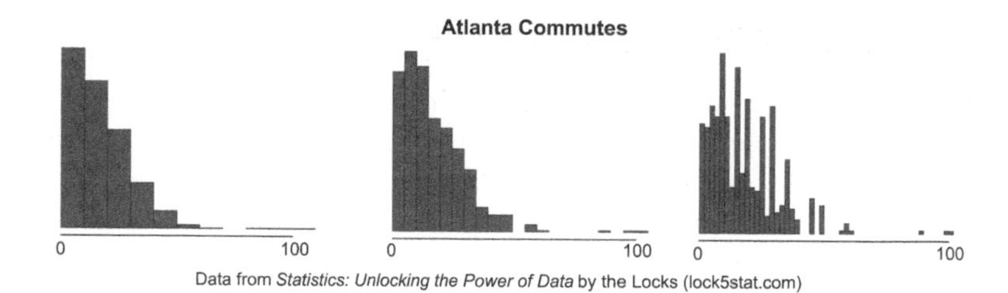

Figure 3.2 Histograms of one continuous variable. We can choose how many bins to use for counting the data. Ten bins (left) seems to lose the detail too much, twenty bins (center) is about right, and fifty bins (right) is so many that it starts to show the heaping effect of people reporting distances rounding off to multiples of five miles, which overemphasizes errors or "noise" in the data.

sometimes a dot plot. This confusion of names is something you will find happening a lot in data visualization. (There's another meaning of dot plot, which appears in Figure 10.4, for example.)

A visualization should always tell the reader what the variables are, preferably by labeling the axes like we did here with "Distance traveled (miles)"; if there are units like miles they should be made clear too. If you find a visualization confuses you because it is not clear what the different visual parameters represent, then the analyst has slipped up.

We also need to make the number of people at each distance clear. In the strip chart / dot plot, the height of each stack of dots shows this without any further effort on our part. If we encoded it to something else, like color, it would not be so easy for readers to compare the number of people traveling different distances.

However, there are limits to this basic chart. If we have a lot of data, we will have a lot of dots, and our strip chart / dot plot will either have to be very tall or the dots will have to be very small. A better format for this same encoding is given by the histogram: chop up the continuous variable into "bins," count how many observations lie in each bin, then draw a bar of height to show the count (Figure 3.2).

When I was a university lecturer in statistics, I always started teaching with histograms, because being able to see and think about the distribution is such a central concept, and it's really

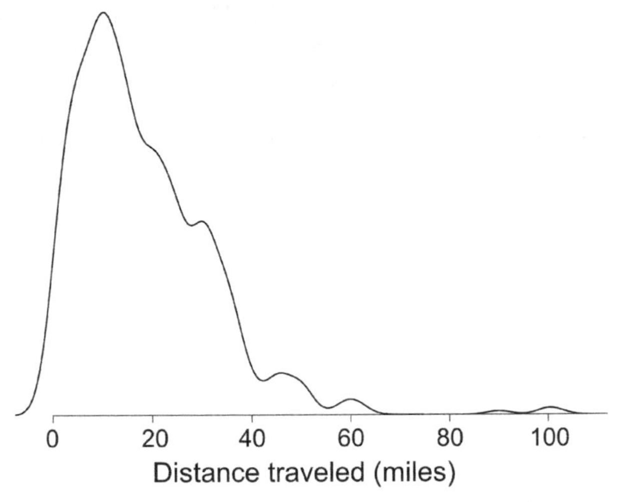

Figure 3.3 A kernel density plot of one continuous variable.

obvious how it is created from the data. It's a springboard to thinking about data in all sorts of more sophisticated ways. The histogram is very widely used and regarded as a standard, basic form of data visualization, but in fact it also turns out to be a crucial concept for today's Big Data problems, which we will encounter in Chapter 15.

It's easy to see how histograms get calculated by counting data, but there's an alternative called the "kernel density plot" that works by replacing each observation by a nice smooth shape (called the kernel) such as the normal distribution (see the box "How to read a distribution"). When the heights of all the kernel curves are added together at each point along the horizontal axis, and the result is plotted, it forms a smooth line that traces out the same shape as we've seen in the histogram.

Kernel density plots are a handy format to use in visualization because there is less "ink" on the page, so they can be absorbed by the reader quickly. On the other hand, they rely on computer power to calculate them, they might not be available from every software package, and they can extend into impossible values because of the smoothing out, which might worry readers (you can see some negative distances in Figure 3.3). Also, the number of observations in the data is lost, and would need to be noted in a caption or nearby for readers to find.

How to read a distribution

Histograms and kernel density plots show us what shape the data take when seen *en masse*. If the data are a sample from a bigger population, then this can give us a clue about the shape the population takes. It is valuable to be able to spot a few characteristics in any chart that shows a distribution:

- Is there one peak or more than one? More than one might suggest the data actually come from a mixture of different populations. Both the distributions shown have one peak (statisticians call them unimodal).
- Is the spread around the central peak symmetric or skewed? How far out does it go? The Atlanta commuting data is skewed (we call this direction right-skewed or positively skewed).
- Are there outlier data that are separated from the bulk of the data by an empty space? There might be something wrong with them – perhaps a data collection error? There might be some outliers in the Atlanta commuting data at 100 miles, which appears as a small bump towards the right. The fact that this is exactly 100 is also suspicious and may indicate heaping.

The same techniques can be used for discrete and continuous variables, the only difference being that we expect the discrete variables to be lumpier in shape as they can only take specific values.

Sometimes, when the distribution is strongly skewed, it can help to transform it. This means that instead of encoding values of the variable to, say, the horizontal location, we encode the values after sticking them through some mathematical function. Logarithms are popular for this: they squash down high numbers more than they do low numbers, so they reduce positive skew.

Any chart using a transformation like this needs to reflect it in the axis, and I think it is probably worth noting it in text too. We don't want to confuse readers by the very trick that we thought would make our work clearer for them. In Figure 3.4, I have encoded the square root of the commuting distance to the horizontal location, and this has made the chart more symmetric.

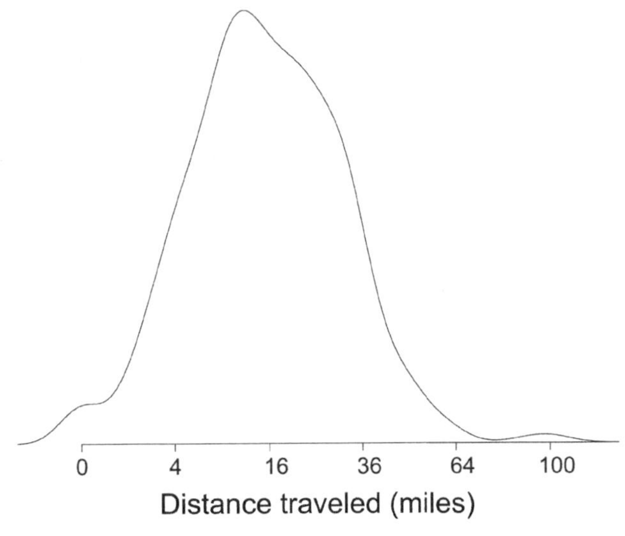

Figure 3.4 Commuting distances in a kernel density plot after square root transformation.

3.2 COMPARING UNMATCHED DATA

Let's say we have some data on one variable and some on another, and we want to compare them. Figure 3.5 has three ways of comparing commute times in Atlanta and New York. Because distance is encoded to the horizontal location, if we want to make comparison easy for the reader, we must arrange multiple visualizations like histograms above each other, and not alongside. Kernel densities, because they are simple curves, can be superimposed effectively. The third option here is a heatmap ribbon, which counts commuters within bins like the histogram, but encodes the count to color. This is compact and can be eye-catching, but color is not suited to detailed examination of the values under the data. In this instance, it might suffice, if the message is simply that many New Yorkers travel a relatively short distance to work.

We can do the same thing when we divide the data into groups and compare the values of one variable across groups. Two histograms stacked above each other allows for some, though not per-

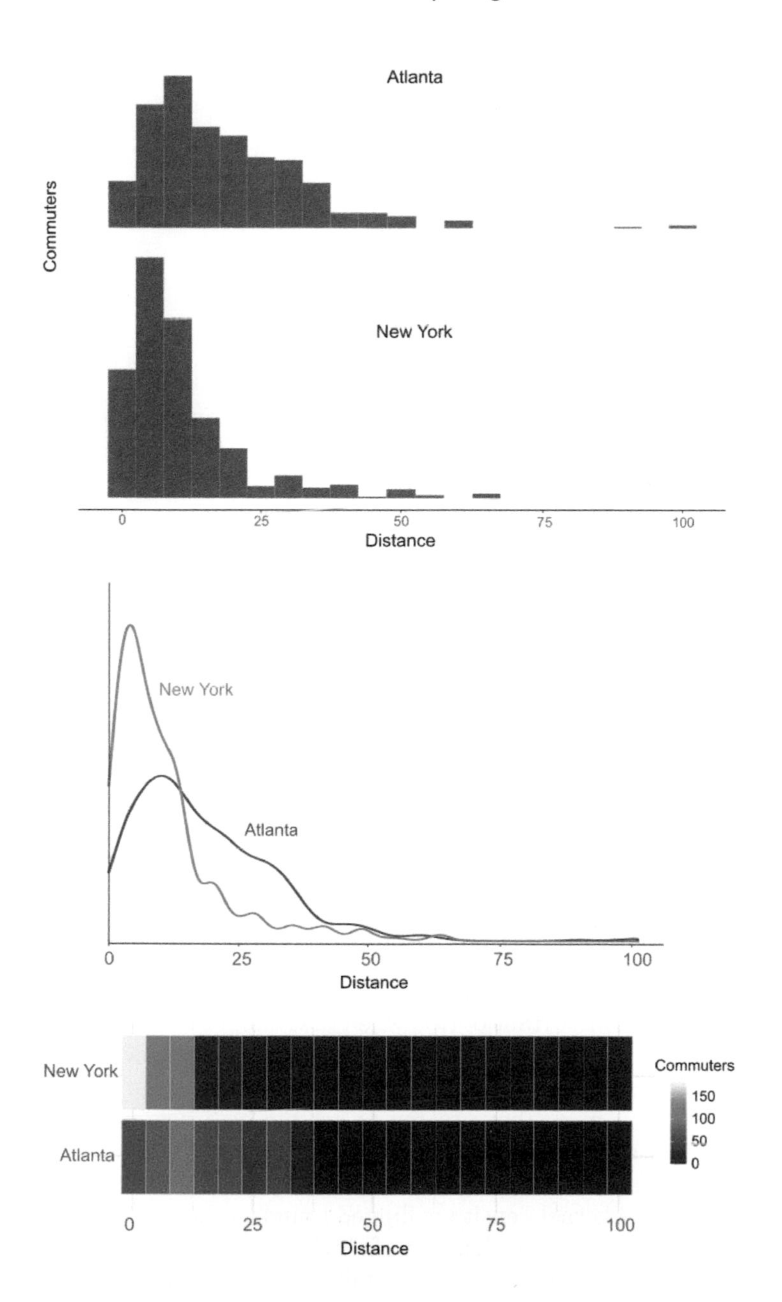

Figure 3.5 Comparing continuous data by a stacked histogram, superimposed kernel density plot, and heatmap ribbons. Atlanta data came from the Locks; New York data are fictitious.

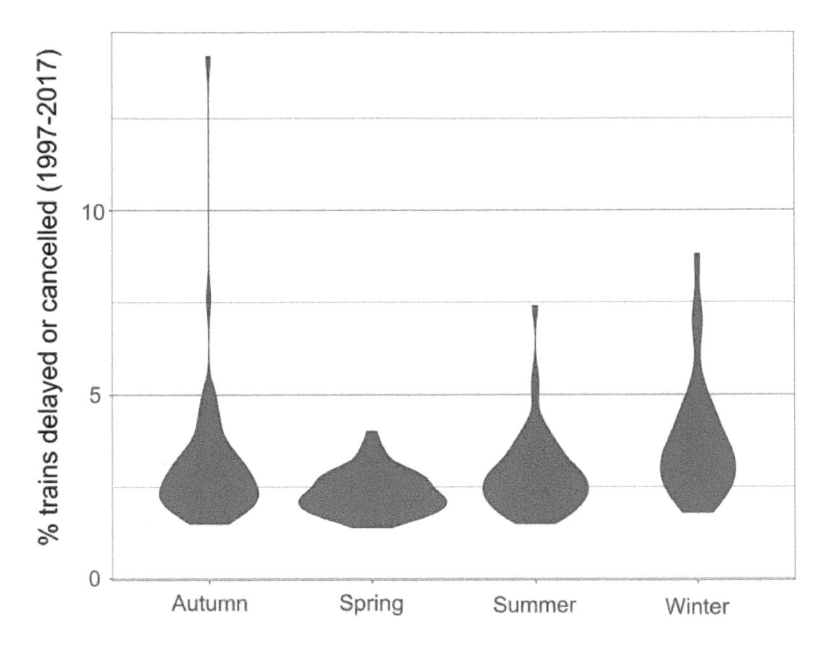

Figure 3.6 A violin plot of the train delay data, comparing four seasons. There is obviously not much difference, except that the winter "violin" is a little higher. I emphasize in the vertical axis label that the data come from 20 years, as a subtle caution to the reader that a lot may have changed in that time.

fect, visual comparison. Two density plots can actually be superimposed, which is clearer. But as the number of groups or variables to be compared grows, the visualization will get too busy.

One really important option when that happens is to stop drawing the data in its entirety and draw some summary statistics instead: we'll deal with this in Chapter 5. If there are a lot of variables, you might have to try to get the best two-dimensional representation you can, and we'll go into that in Chapter 12. But for now, a variant on the kernel density plot, called the violin plot, can help. The kernel density is flipped round to the vertical dimension, and mirrored on either side of a line (Figure 3.6). Because this is quite compact, you can line up quite a lot of groups or variables before it gets overwhelming for the reader.

Long, thin visualizations for each group can also be stacked above one another. This works well with heatmap ribbons, and when done with kernel densities it can look quite physical, like a three-dimensional landscape, which helps readers absorb the information quickly (Figure 3.7).

If you have to compare several variables over several groups, you will need multiple charts. Ask yourself whether the reader will typically compare groups or variables. To continue our example, people may be interested to compare commute times between Atlanta and New York, and work times likewise, but there's less interest (we imagine) in comparing commute time and work time in any one city. So, we would make a chart – perhaps a violin plot – of commute times in various cities, and another of work times in various cities, and so on for different daily activities. Sometimes, you just have to compromise with complex, data-heavy visualizations.

3.3 COMPARING MATCHED DATA

When we are comparing two variables or two groups in the data, and we can also link the individual observations together across the comparison, we have matched data. The most common reason for this is having data on the same people (or whatever you are measuring) at two timepoints. The charts in the previous section are applicable, but they omit information: the link between data points.

It's important here to compare like with like. Obviously, we should match the correct data together, and for the same reason, it will be confusing if some of the data are matched and some are not. Suppose you are surveying customers on how much they spent on groceries in the last month. Your company ran this survey in 2015 with a small sample of about a hundred customers, and now the boss wants to know if things have changed. One option would be to run a whole new survey. It's very unlikely that anyone from the 2015 survey will be included, and you would then have unmatched data.

However, this nagging doubt will stay with you: if you see a difference, is it that shopping habits have changed generally, or is it just bad luck that you got a different batch of people? There are statistical ways of weighing up the evidence for these competing explanations, mainly by establishing how unlikely the second (bad luck) one is, but you could improve your information by getting the list of people who took part in 2015 and contacting them again. If they have all started spending more, or less, as individuals, then

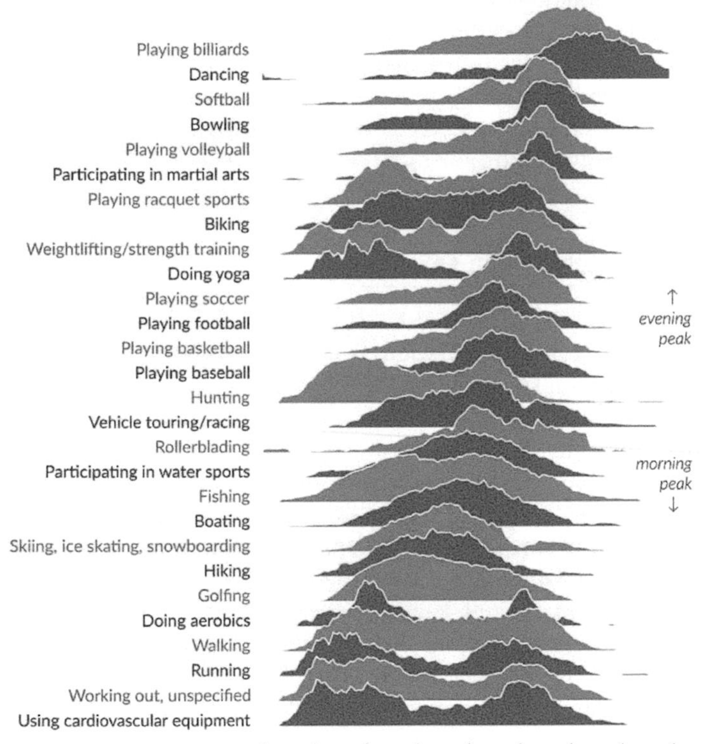

Figure 3.7 Stacked densities of time of day for various leisure activities (note that no kernel has been used, so there is no smoothing of the curves). A "joyplot" by Henrik Lindberg, named after the sleeve of the Joy Division album *Unknown Pleasures* which reproduced a visualization of astronomical data by Harold Craft in 1970, then a student at Cornell University. Reproduced with permission.

that would be stronger evidence for a widespread shift in habits, or disposable income, or something else.

Visualizing this requires either drawing all the data with the matches, or calculating differences and drawing those instead. Each person in the customer survey could have a 2015 spend and a 2017 spend, and a line or other device joining them. Or, they could just have the change (positive or negative) from 2015 shown. It would amount to the same thing but the first option shows the reader the distribution at each timepoint. However, there's nothing to stop you from doing this in a separate, scene-setting chart before you get down to how much each person has changed.

One good way of showing these individual-level changes is by a line chart where each observation has its own line and time (or whatever the matching is) is encoded to the horizontal position (Figure 3.8). This is a case of small multiples, which is a general concept that reappears in several places in this book. There are multiple small charts and each one shows a different aspect of the data. Comparisons can be drawn while not losing sight (literally) of the bigger picture. If a really data-heavy set of small multiples is going to be viewed online, then it can be made interactive, so readers can pick what they want to compare (Chapter 14).

Another option that is clearer for large data sets, but not so intuitive for readers unaccustomed to statistics, is to draw a scatter plot, like we saw in Chapter 2. This could have the 2015 spend on the horizontal axis and the 2017 spend on the vertical, in which case a line of equality should be drawn, on which the two spends are the same, above which 2017 is greater, and below which 2015 is greater. Another option is to encode the 2017 values to the horizontal, and the change to the vertical. Either way, this will need some accompanying text to introduce it to the audience.

Or you could divide up the change into ordered categories, maybe "Notably less," "About the same," and "Notably more," and encode those as an additional parameter, maybe color. There are plenty of other options too. Sometimes, with two variables and their respective changes, it's possible to draw a scatter plot and have an arrow for each data point that points in the direction of the change, essentially showing that it moved from *this* x and y to *that* x and y. This could be horribly cluttered, but when the changes have a clear consistent pattern, it can be very effective. We will encounter one in Figure 9.2.

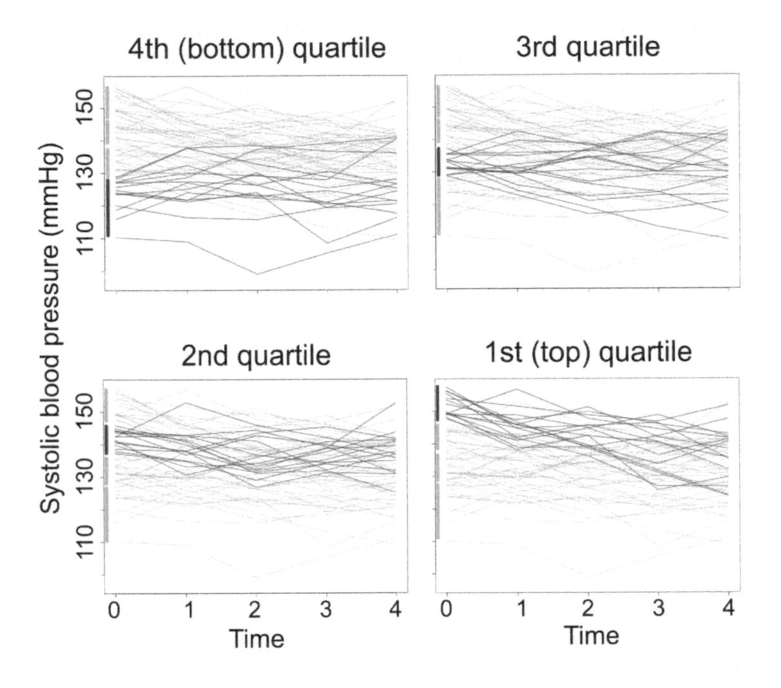

Figure 3.8 A line chart used to show matched data for a fictitious study of blood pressure treatment. To tackle overcrowding, I used small multiples, replicating the chart four times, faintly, and highlighting a different part (quartile, see Chapter 5) of the data each time. Now we can see that it is the people with the highest blood pressure at the beginning of the study (bottom right) who benefitted the most, while others did not consistently change.

3.4 ASSOCIATIONS

The idea of associations between two variables is a really important one throughout statistics and data science. Suppose we notice that the customers who have been customers of our company the longest are the ones who tend to be spending more on their groceries. Now we are talking about combinations of one variable (time with company) with another (change in groceries spend), and we can do this more generally, to find out whether the value of one variable is associated with another.

This can lend evidence to a hypothesis that our company attracts people earning higher-than-average incomes, who spend more (good news for the boss, perhaps). But it might be the reverse: those who can't afford their groceries are economizing by no longer being our customers (the same data have just become bad news). Or it might be that both have a common cause: young people can't have been with us that long, and they don't earn as much and don't have as many children, so don't spend as much on their groceries (no interesting news at all). Or there may be no cause-and-effect connection we can understand at all, but even so, we can use time with our company to predict people's groceries spend: a prediction, not an explanation.

The mainstay of associations in discrete or continuous data is the scatter plot. A common problem is that markers might coincide in one spot, when observations share the same values. One way around this is to make the marker thicker or darker at a shared point, but it is hard for the reader to judge just how many points there are from the size or color of the marker, because these are visual characteristics which are not accurately perceived.

Another approach is to jitter them by adding small random numbers to both variables, which will spread the markers out into a cloud of dots around their common location and give a general impression of the number of data points. Of course, we should tell the reader that we are jittering in accompanying text. We revisit jittering in Chapter 7.

Important features to look for in a scatter plot are whether there is one cloud of dots or several clusters, whether there is an upward or downward slope to the cloud of dots (indicating correlation, which we return to in Chapter 6), and whether there is any curvature to the slope. Remember this old statistician's saying: **correlation is not causation**. Just because one variable goes up or down as the other goes up or down does not mean one affects the other in any way.

Sometimes, the individual variables encoded as horizontal and vertical can be shown in their own right, by having histograms or density plots in the margins of the scatter plot (see Figure 3.9). Another option is to have a short line in the margin where each data point occurs, which is called a "marginal barcode," or sometimes a "rug plot." The concept lends its name to an important idea in data visualization: "marginal distributions" are for one variable at a time, "joint distributions" are the scatter plot itself, and we previously saw some "conditional distributions" where we look at

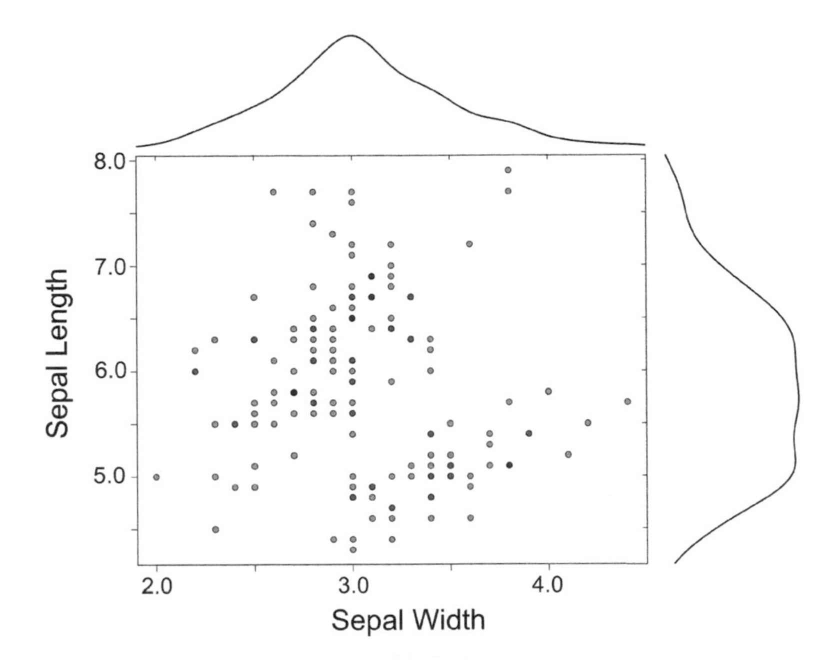

Figure 3.9 A scatter plot of iris flower shapes with marginal kernel density plots. Neither marginal density on its own would lead you to suspect that the data had contributions from more than one species, but the scatter plot, visualizing them jointly, shows it.

the distribution in a certain subset of the data (conditional on belonging to that group).

Other variables can be encoded to parameters affecting the markers. Color is a good option for categorical variables, as long as there are not so many that it overwhelms the reader like a kaleidoscope! A popular approach is the bubble chart, where the size of markers is a third variable. Recall, though, that area is not well perceived, so you should be prepared to accept the third variable as ordinal at most. The bubbles can also obscure one another if they are opaque. We will take a look at how to show many variables at the same time in Chapter 12.

Although maps contain horizontal and vertical positions, they are perceived in their own way, accelerated by the familiarity of the location. Maps deserve at least a chapter of their own (Chapter 13).

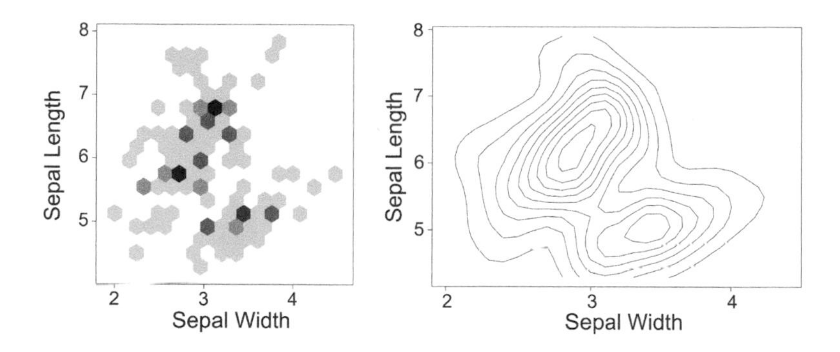

Figure 3.10 Alternatives to scatter plots: hexagonal binning (left) counts the observations (flowers) inside each hexagonal area and encodes the count to color, with darker hexagons containing more observations, while the contour plot (right) adds together two-dimensional kernel densities around each observation and then draws lines joining together equal density points.

If the scatter plot is a two-dimensional dot plot, you might wonder what the two-dimensional histogram and density plot look like. In the same way that histograms count data in bins, we can subdivide the two-dimensional surface into bins and count within them. A square grid works, but hexagons create a smoother effect by avoiding the sharp corners. Recently, hexagonal binning or "hexbin" plots superimposed on maps have become very popular (Figure 3.10, left).

We can also smooth the density over the two-dimensional surface and show this as though it was the height rising out of the page. Contour plots are a simple way to do this. Although the density above a two-dimensional surface can't be drawn directly (short of using a 3-D printer), there are a couple of options we can use: we can take slices through the joint distribution and then draw conditional distributions, mimicking a 3-D elevation seen from an angle (in the style of Figures 3.7 and 12.3), or we can draw contour lines in the same way that maps show elevation above sea level (Figure 3.10, right).

Two cautions need to be given. First, continuous or discrete data can be made to look like there is a bigger or smaller difference

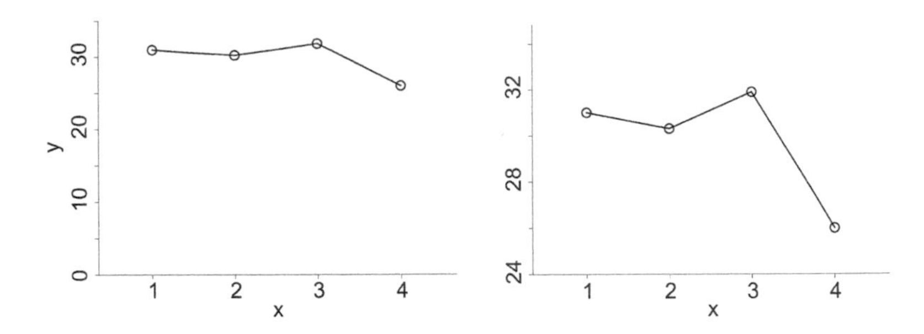

Figure 3.11 Use and abuse of axes: sometimes, it is misleading not to show the whole range that a variable *could* take.

than there really is by zooming in on the area the data occupy. You can see this in Figure 3.11, where the left image shows a variable that extends down to zero and the right image zooms in on just the available observations. If the data showed sales dropping in a company, it might be tempting to show the left one. If it showed crime rates falling for a police force, the right one might be preferred. Because of this, you may sometimes hear a very strict rule that all axes should begin at zero. That misses the point; you should show the whole picture and try not to mislead the reader. As a counter-example, it would be silly to extend the axis representing the year in the train delay data down to zero.

The second caution is that you may sometimes see broken axes like that in Figure 3.12 (right). In fact, the left and right images show exactly the same data. You might not notice the small symbol and the change in numbers on the vertical axis in the right image, and it could easily mislead readers. There is no reason ever to use a broken axis like this.

Secondly, some numbers, like countries' GDP in billions of US dollars, are totals which would make more sense divided by the population (per capita). The problem is that the observations in the data are quite different: some are bigger than others, so it is hard to compare like with like. Similarly, the number of times some event has happened should be shown as a rate if different observations have been followed for different lengths of time. You can imagine how the customer survey above would be undermined if some people replied with how much they had spent on groceries in

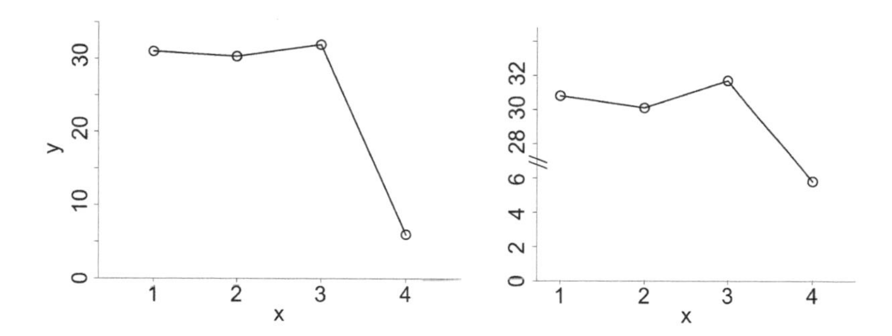

Figure 3.12 Use and abuse of axes: breaking the axis is misleading.

a month, and others in a year: they would all have to be converted to a *per month* figure.

This chapter has been all about showing data. We will address the statistical summaries in Chapter 5 and predictive models in Chapters 9, 10, and 11.

CHAPTER 4

Percentages and risks

WHEN THE DATA are just a list of "yes" and "no" for each observation, the options for how you represent it statistically are seriously limited. That's not to say that data like this is not useful, it's just simple. What we are really dealing with here is membership of a category, and "yes" and "no" are the simplest categories. Whenever you fill out a survey and tick a box, you put yourself into a category. Maybe there are only two options: ticked or not ticked, or maybe there are multiple categories. If you tell an election pollster how you intend to vote, you fall into one of several categories.

Sometimes, questionnaires ask people to "tick all that apply" (for example: "Which of these software packages do you regularly use?" followed by a long list), so they can fall in more than one category at the same time. Some categories go in a specific order, for example "How do you feel about our service? Very unhappy – Unhappy – Happy – Very happy." And finally, some categories are nested inside others: you might work in the Data Analytics Team, which is part of the Marketing Department, which is part of the Commercial Services Division of a company. In each case, there are some helpful visualization formats.

4.1 SHOWING ONE VARIABLE AT A TIME

To describe a variable with two categories, like "yes" and "no," only three numbers are needed: how many said "yes," how many answered the question, and what percentage that is. If you asked 200 people and 60 said "yes," that's $60/200 = 30\%$. Occasionally, you might see people using proportions, where you just divide the number saying "yes" by the number who answered, so 30%, written as a proportion, is 0.3. Whichever you use, visualizations will look the same except for the labels on the axes, but most readers will be more accustomed to seeing percentages.

What about the people who said "no"? Their numbers are implicitly $200 - 60 = 140$, so you don't have to write them out. But, if there were some "yes" answers, some "no," and some left blank, then the blanks should be acknowledged somehow. You could treat it as three categories ("60 (30%) said yes, 110 (55%) said no, and 30 (15%) did not answer"). Alternatively, state how many were blank and then reduce the total accordingly to just the "yes" and "no" answers: "$60/180 = 33\%$ said yes, while 20 did not answer the question." In visualization, you are often drawing the percentage, so make sure that the count in each category is not lost: include it in a label or accompanying text.

The count matters because small numbers cause the percentage to be less stable. With only 2 observations, for example, the percentage has to be 0, or 50, or 100, which seems much more dramatic a change than it really is, unless you can see that it is based on small numbers. I'll talk about data versus statistics much more in Chapter 5, and how to show uncertainty in the stats in Chapter 8.

Simple percentages are best encoded as lengths, for example a bar chart or dot plot. To be a little more eye-catching, you could use a pictogram (see Figure 2.8), which is also encoded to length if you line up the icons. To give an impression of the whole 100%, the length can be shown against a 100% background (Figure 4.1 (left)), or to compare with some targets or other reference values, a variation on this is called the bullet chart (Figure 4.1 (right)).

When there are more than two categories, they should all be visible or listed in accompanying text, even just to show that nobody chose it. For example, customers buying goods online can be classified according to their home country (where the billing address is); there will be about 200 categories but each person can only fit in one of them. One bar or length will no longer suffice, and sometimes there might be a big contrast between one percentage (say, the number of customers from the United States) and another (the number from Swaziland). This will make the Swazi bar so small that it will be impossible to see or judge its size.

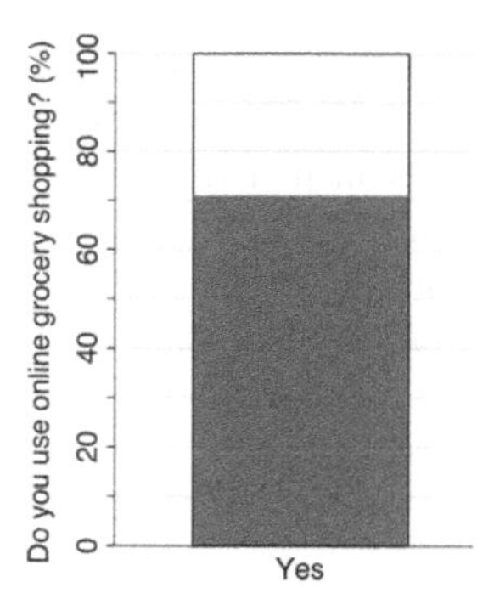

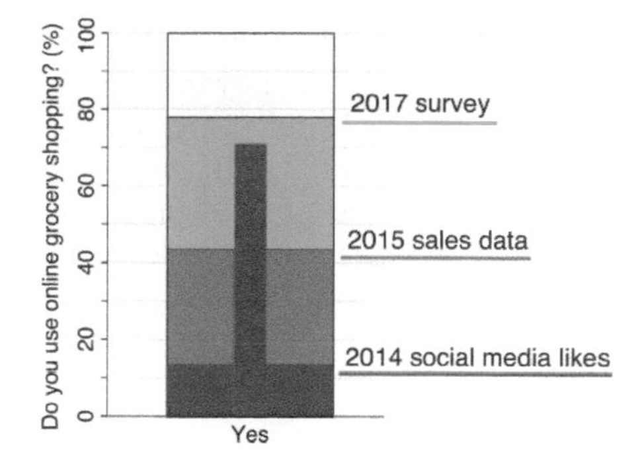

Figure 4.1 Percentage answers to one binary question.

We previously explored the problems with pie charts, but we could consider some kind of waffle to get around this problem (Figure 4.2). Of course, small categories can be lumped together as "Other" and a footnote can explain what it contains. When there are large numbers of categories, readers will find it hard to absorb all the information, and it might be best to collapse some together. User testing can help with finding the right threshold for this.

Ternary plots are a way of simply showing any variable that contains three categories – but no more than three (Figure 4.3). The abundance of each must add together to 100%. They are popular in geology, chemistry and food science – all areas where people deal with mixtures. At each corner of the triangle, we find 100% or one of the categories and 0% of the others. In Figure 4.3, a day that occupied the top corner of the triangle would comprise all work, no play and no rest.

There are two special cases to think about. First, "tick all that apply" questions, where the percentages may well add to more than 100%. Essentially, this sort of question has to be broken apart into a series of binary yes/no questions for each of the options. Second, ordinal variables have an innate order, and that always has to be respected in visualization. Unfortunately, this introduces another restriction into making the visualization, but compromise is inevitable.

4.2 COMPARING UNMATCHED DATA

If you are interested in comparing the percentages arising from two variables, encoding them as lengths and presenting them next to

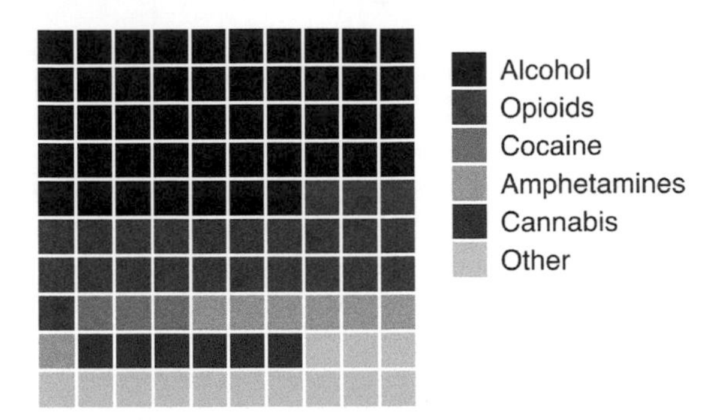

Figure 4.2 A 10-by-10 waffle shows percentages from six categories with one highlighted. Data from "The Global Epidemiology and Contribution of Cannabis Use and Dependence to the Global Burden of Disease: Results from the GBD 2010 Study" by Louisa Degenhardt and colleagues.

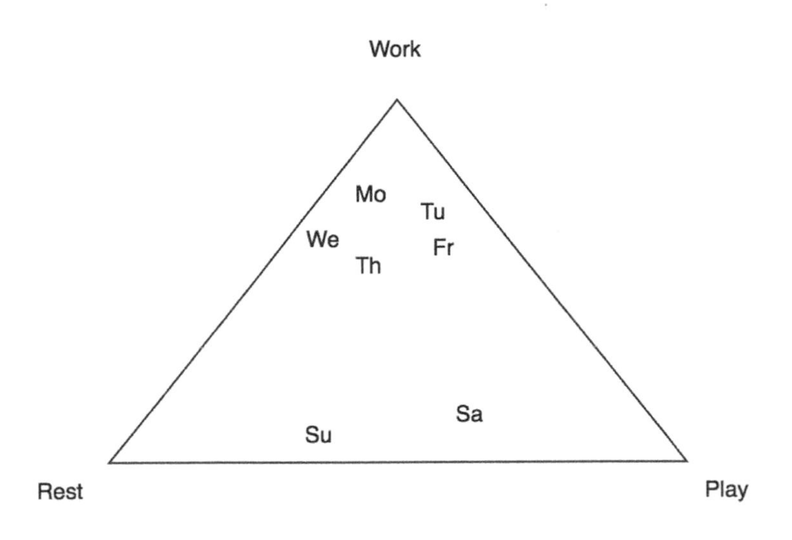

Figure 4.3 A ternary plot of proportions of activities for each day of the week

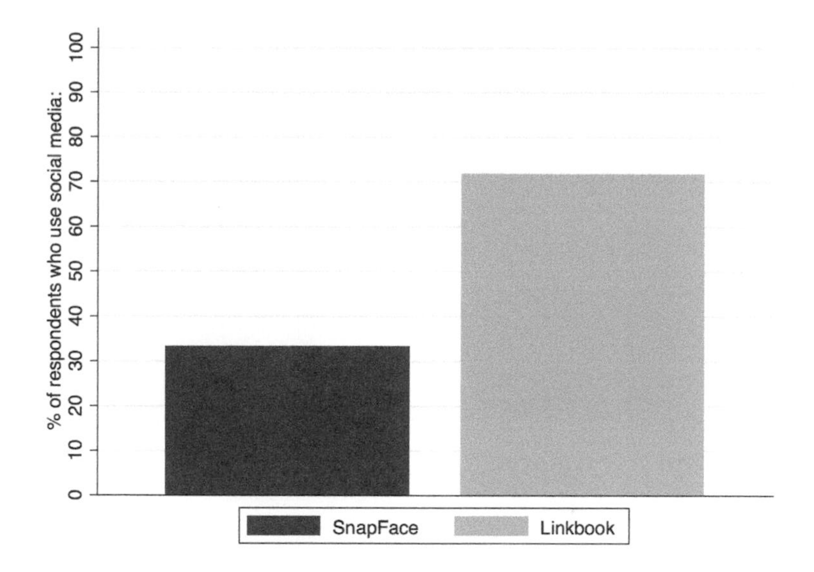

Figure 4.4 Comparing percentage answers to two questions

one another is the best first choice (part of an imaginary customer survey is in Figure 4.4). Areas, even the waffle plot, can't be visually compared without some effort or guesswork on the part of the reader. You might feel tempted to add the actual percentages as text to a chart like this, but there is no need if the axes are labeled clearly and the encoding is appropriate; any extra text will only clutter the image and distract the reader. These percentages represent unmatched data because we are not explicitly linking one person's reply in one variable to the same person's reply in another variable. If they differ, that's all there is to say about it.

If there are many variables or categories to be compared over many timepoints, you can line them all up alongside one another, but comparison gets harder as they move farther apart. Important comparisons should appear close together, and to do this, there might have to be compromises, for example, show all the categories in one chart and then the important comparisons only in another, or collapse less crucial categories together. If only one of the categories matters, why not collapse it to a binary yes/no, which can be shown in a very compact form?

The option of stacking a bar chart (Figure 4.5) is given in many spreadsheets and data analysis software packages, but this is prob-

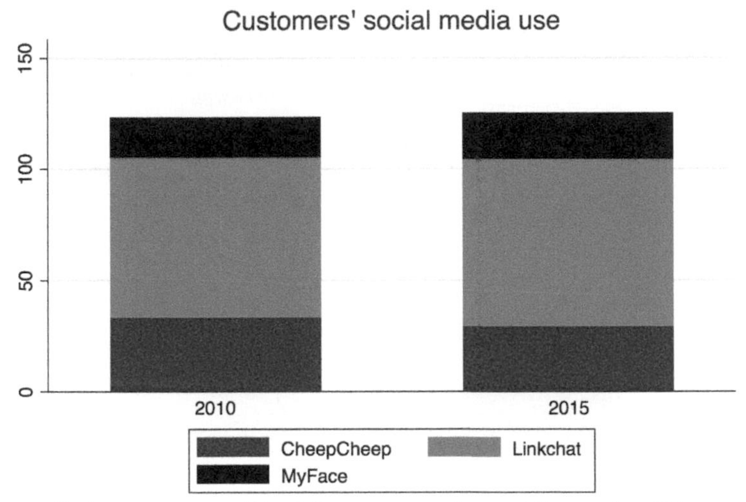

Figure 4.5 A stacked bar chart; compare with Figure 4.6

lematic. Although the category at the bottom of the bar can be visually compared across the bars, all the others on top of it cannot (without using a tape measure...) because of the problem shown in Figure 2.7: they don't start at the same level.

4.3 COMPARING MATCHED DATA

As with continuous and discrete variables in Chapter 3, we sometimes have matched data. The clustered bar chart, or its dot plot and line chart equivalents, is key here (Figure 4.6). You could help readers to also see which question has the highest percentage, or the biggest change, by arranging the pairs in that order (Figure 4.7). This is a simple example of emphasizing the message without distorting the facts.

When comparing timepoints, compare like with like. For example, if the respondents to the 2017 survey were very different from the 2014 survey, you need to tell the readers about that in a footnote, or think twice about comparing them at all. It might be better to compare only a subset that took part in both. Sometimes, the matching of data is done in some way other than time. A medical study of arthritis, for example, might compare the affected hip

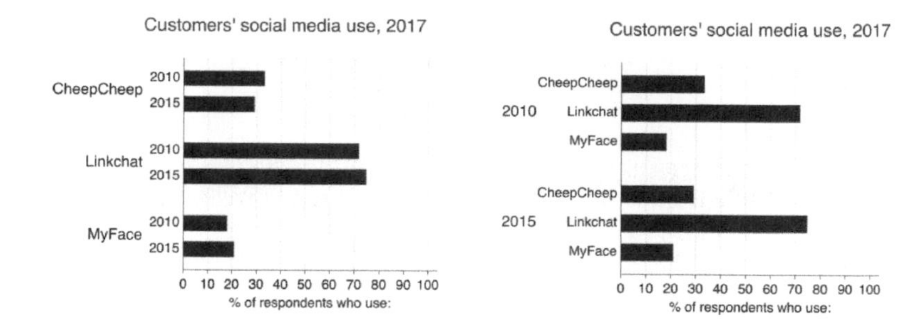

Figure 4.6 Clustered bar charts comparing three binary variables over time; notice how the small changes are easier to see in the left chart, and easier than in the stacked version above.

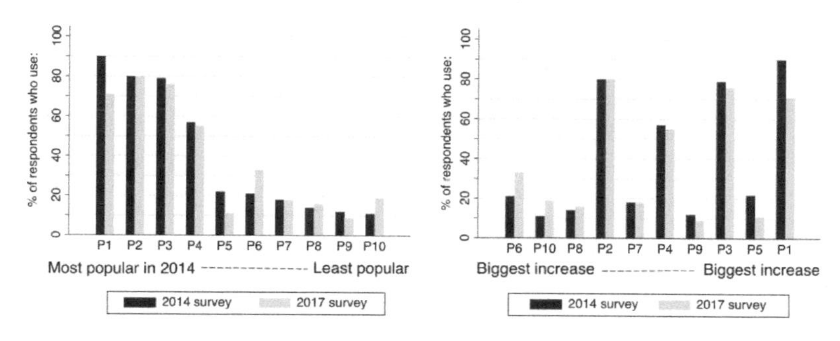

Figure 4.7 Comparing ten binary variables over time, ranked by value at the first timepoint (left) and the change (right).

to the healthy hip in each participant. It's still matched and needs to be shown as such.

Sometimes, what matters most is not that so many people gave a certain answer in 2014 and so many in 2017, but rather how many changed one way and how many the other way. This takes us away from the raw percentages and into ratios and differences – statistics that measure the change itself. We'll look at them in Chapter 6. But consider Figure 4.7 for now. Has everyone's response to question P2 stayed the same? Or have the same number gone from yes to no as went the other way? We can't tell from a bar chart like this. One way of showing these flows of matched data is with a parallel sets plot (Figure 4.8), though they can become very cluttered quite quickly; this is analogous to the matched line charts we saw in Figure 3.8.

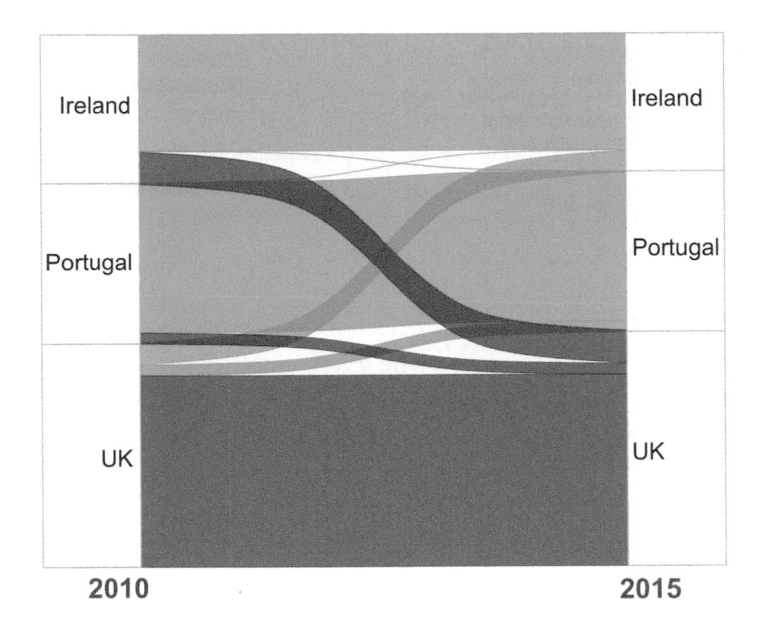

Figure 4.8 Showing movement between categories in matched data with a parallel sets chart – 2 people have moved from Ireland to the UK, 1 from Ireland to Portugal, 2 from UK to Ireland, and 1 from Portugal to Ireland.

4.4 CATEGORIES WITHIN CATEGORIES

When categories are nested inside one another, the tree format comes into its own. A decision tree can be linear, with the branches always heading down (or up, though down seems more common) the page as the data get subdivided (Figure 4.10). It can also be radial, so that concentric rings show the categories and the data are subdivided more when moving away from the center. A popular form of this is the donut chart, which like a pie, encodes proportions as angles (and therefore areas); further rings can then contain subdivisions quite intuitively and this is sometimes called a sunburst chart. The treemap is another option (Figure 4.9). All of these suffer from the problem that we might have to compare non-adjacent items, and they might not have the same starting point (because the location of the branch dictates it). The treemap and donut chart also rely on perception of areas.

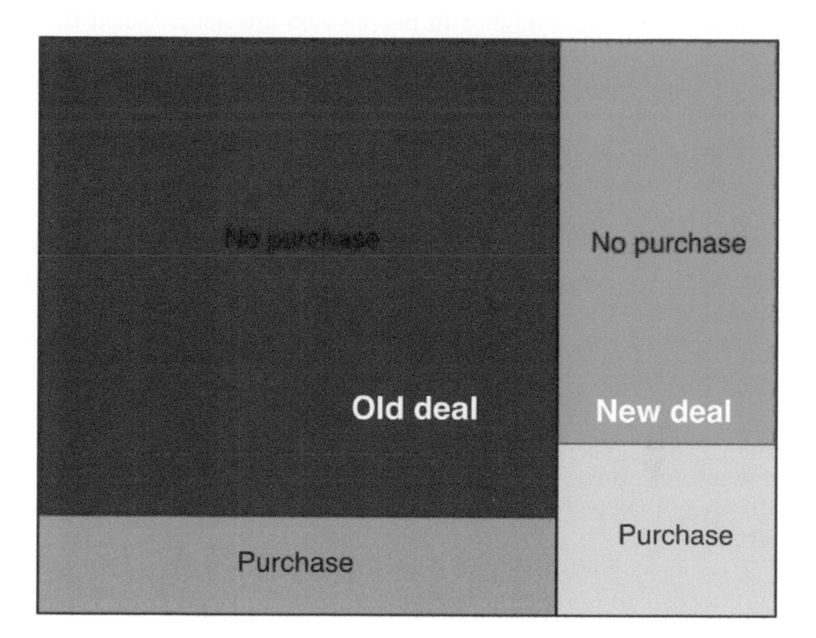

Figure 4.9 A treemap: a marketing experiment compares two deals and whether website visitors make a purchase. Although a minority of visitors were offered the new deal, the higher split of that green segment into purchases and no purchases shows that it was more popular than the old deal.

4.5 ASSOCIATIONS

The idea of associations between two variables is a really important one throughout statistics and data science. So far, I've shown you some ways of comparing percentages over categories and sometimes comparing categories over timepoints. That means combinations of one variable (the categories) with another (time), and we can do this more generally, to find out whether the value of one variable is associated with another.

For example, imagine data provided by people receiving drug treatments for arthritis. They are asked how satisfied they are with the pain relief they feel they are getting from the treatment ("not satisfied" or "satisfied") and also whether they have tried any complementary therapies, like acupuncture. Perhaps the percentage trying acupuncture is higher in people who are not satisfied than in people who are satisfied. This could lend evidence to a hypothesis that inadequate pain relief leads to people seeking their own additional treatments in the form of acupuncture.

It might be the reverse: the experience of receiving the acupuncture with unhurried one-to-one care makes them less satisfied with the short medical consultation. Or it might be that both have a common cause: some people just don't trust conventional drug treatments. Or there may be no cause-and-effect connection we can understand at all, but even so we can use lack of satisfaction to identify people who might want to try acupuncture, as a prediction not an explanation.

In cases like this, we need to look at conditional percentages: the percentage trying acupuncture in the "not satisfied" group (or we could call this "conditional on satisfaction"). For conditional percentages, there are several options in addition to clustered bar charts.

Treemaps divide up a rectangular area on the basis of one variable encoded to horizontal length and another to vertical (Figure 4.9); although we know that areas are not accurately perceived, the lengths on either side are.

Decision trees show the breakdown of the data by one variable then another in a very intuitive way, though they are generally just diagrams that don't actually encode data visually. However, pictograms or waffles can be superimposed at each of the points where the decision tree branches or terminates, giving at a glance a feel for the numbers and how they propagate through the tree (Figure 4.10). The numbers can be real observations or hypothetical ones: a tech company's marketing department might show various scenarios where users move from visiting their website to downloading their app to making in-app purchases. Chapter 10 will look at visualizing predictions as a result of statistical modeling like this, and Chapter 11 will look at decision trees in more detail.

Ideally, the reader should be able to mentally flip the percentages around to whatever way interests them. We have shown how many website visitors offered a particular deal made a purchase, but if they wanted to know how many of those who made a purchase had been offered a particular deal, this should be easy to do visually, even without counting the squares of the waffles or any more than negligible effort. Reversing conditional proportions like this is a common problem in statistics.

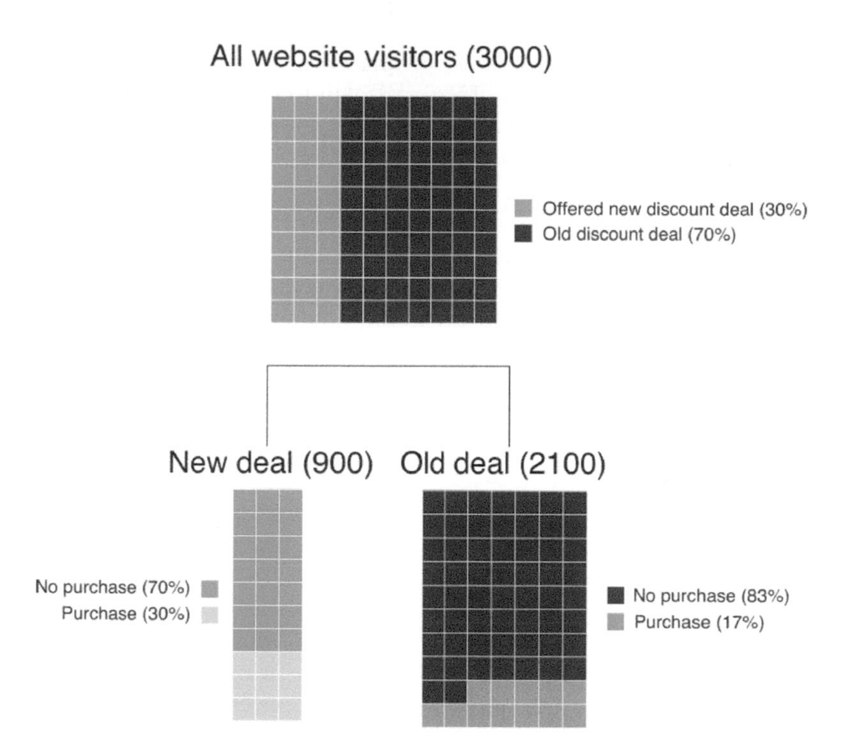

Figure 4.10 A decision tree with waffles: a marketing experiment compares two deals and whether website visitors make a purchase.

The explanation is not something you can get from the data: it has to be discussed and considered by experts and the arthritis patients. Qualitative research is used in many settings, scientific and commercial, to back up the patterns in the data with insights like these. The relevance to data visualization is that we are always conveying a message to some extent, and in the case of associations between variables, that message is sometimes a step removed from the data itself. If you are making visualizations, be careful not to impose your own interpretation too much when showing associations. If you are reading them, don't assume that the message accompanying the data is as sound and scientifically based as the data themselves.

This chapter has been all about showing the data (or that simplest of statistics, the percentage). We will look at how to visualize predictive models for future data in Chapter 10.

4.6 RISKS, RATES AND ODDS

We often want to show the risk of something happening, and this is just a proportion or percentage like any other, but we have to be careful to be absolutely clear about the denominator, either in the image or in accompanying text.

What happens when the risk is no longer a fair comparison? If the marketing department we just considered had tracked customers over a long time and counted whether someone had made any purchase or no purchases, then it might not be a fair comparison to look at long-standing visitors in the same way as newcomers: surely the newcomer is less likely to make a purchase as they have had less time to do so.

In cases like this, we should divide the number of purchases by the length of time someone has been visiting our website (or how many times they have visited). This gives us a rate, and although rates do not fall into the range 0% to 100%, we can visualize them using the same techniques we've been looking at in this chapter.

Sometimes, for computational reasons, we have to talk about the odds of an event happening and not the risk. The odds is just the number of "yes" (or equivalent) answers divided by the number of "no" answers (not the total). In the marketing example above, the odds of making a purchase if you were offered the new deal is $270/630 = 0.43$. Odds are never written as percentages (43%) to avoid confusion. Try not to use odds if you are making data visualizations, as it has been shown that people find them more confusing than risks and other straightforward percentages. We will encounter the odds again in Chapter 6, including ways to avoid it.

CHAPTER 5

Showing data or statistics

EVERY TIME WE SUMMARIZE data, we are representing them by a statistic. The mean is a statistic that tells us where the middle of the data is by adding them all up and dividing by the number of observations, and likewise the median is a statistic: the value that has half the data above (or equal to) it and half below (or equal to) it. The percentage of data falling in a category is also a statistic. Technically, even the counts shown in the height of a histogram's bars are statistics.

There is a choice between showing the individual data and the statistics, and the goal should be clarity. It is nice to see all the data so the reader knows that nothing is being hidden, and can explore the data to answer their own questions, but that can become too busy to be helpful for many readers who want a headline. Returning to the commuting data from Chapter 3, if we wanted to compare several cities, then showing stacked histograms or superimposed kernel density plots, as we did for two cities, would just be information overload.

The headline message might be adequately captured by just showing a summary statistic for each city (Figure 5.1). If we show just a marker for each statistic, this is called a dot plot (there is a different visualization in Chapter 3 with this name, but that is quite rare in practice).

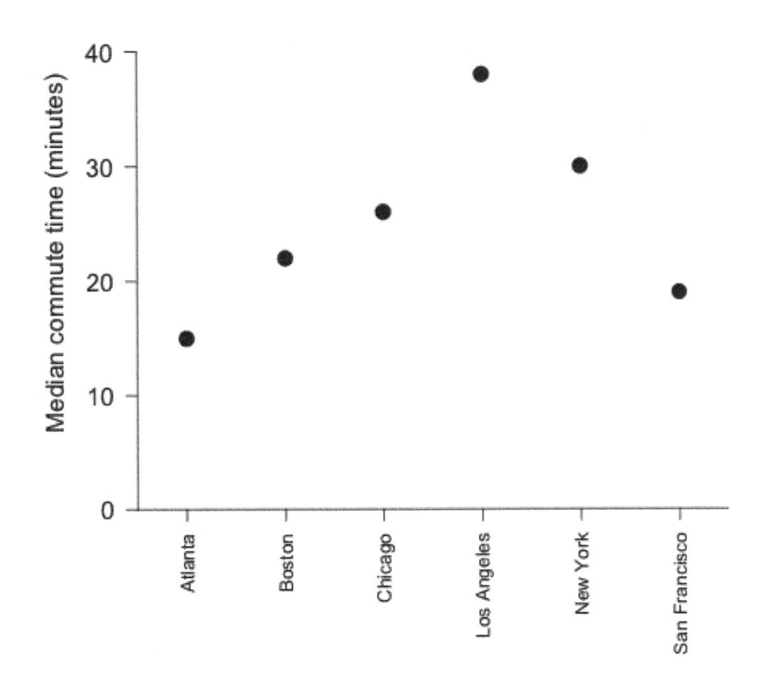

Figure 5.1 Dot plots show just the mean (or another statistic) for a series of groups in the data, or variables. Dot plots can also have the statistic encoded as the horizontal position, for example Figure 10.4. Fictitious statistics, except for Atlanta; see Chapter 3.

5.1 CHOOSING DATA OR STATISTICS

Apart from clarity, the visualization should take into account the readers' needs and the subject's anonymity. If the reader just needs to know a statistic, for example the median house price in various neighborhoods, don't waste their time with each house. Interactive online visualizations allow for some flexibility in this (Chapter 15). The creator of the data visualization might also need to work within the confines of agreements to collect or use the data, or wider laws. Perhaps the individual who has participated in some research must not be identified, even accidentally.

Publishing only statistics would help here, though we may still have to redact some of them, such as the earnings and demographics in various groups if one of the groups has only one person in it; in this case, publishing statistics will be as good as publishing that person's data. The risk of re-identification is never completely removed but it can be managed.

You can also combine the statistics with the data, as long as it doesn't get cluttered. Biostatistician Frank Harrell has suggested spike histograms, which are histograms with little spikes extending down from the horizontal axis that show the locations of the mean, quartiles or whatever statistics you might want to show. They are compact and allow the reader to choose what to look at.

It's critically important that all the relevant statistics are shown, not just the ones that look exciting. The more we indulge in cherry-picking some results and ignoring others, the greater risk we run that those statistics will turn out to have been nothing more than chance fluctuations, which will disappear if we repeat the data collection.

This is perhaps the single most damaging problem with how statistical analysis is practiced today. It has different names, like multiplicity, p-hacking or the garden of forking paths. It is concerning in visualization because we have no choice but to curate information and it takes an experienced analyst to know when this crosses over into hiding some of the facts. We'll encounter a real-life example of this in Chapter 6.

5.2 THE STANDARD DEVIATION

The mean has an accompanying measure, called the standard deviation, of how spread out the values are around it. Unfortunately, its interpretation is not very intuitive. The normal distribution in Figure 5.2 is often encountered in continuous variables, and if your data have a histogram like this, the mean will be at the highest point, and the standard deviation is the horizontal distance from the mean to the point where the histogram bars change from bulging upwards to bulging downwards. Frankly, even a statistician finds this dull.

So, a more intuitive way to express it in words is that 95% of the data should lie between two standard deviations below the mean and two standard deviations above the mean. Remember that

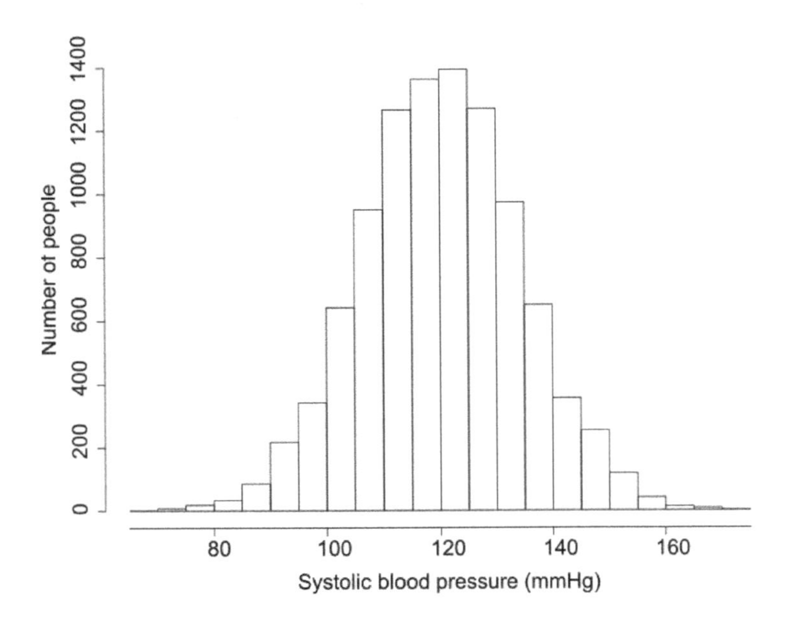

Figure 5.2 The normal distribution is a commonly occurring shape in many different settings.

this applies for normally distributed data, not with other shapes. This is sometimes called a 95% reference range. You can show it on visualizations by an error bar extending above and below the mean (Figure 5.3). But, as we will see in Chapter 8, the error bar can mean other things too, so we should always state what it is in an annotation.

In two dimensions, such as we've seen in scatter plots, you can have a reference range for the horizontal variable, and another for the vertical, and visualize this as two sets of error bars at right angles, or as an ellipse. The cross where the error bars meet, or the center of the ellipse, will be at the point defined by the two means. However, although the means tell us where to center the ellipse, and the two standard deviations how elongated it should be in the two dimensions, we don't know if it should be tilted at an angle or not. That is the role of the correlation, which completes the picture and we'll discuss in Chapter 6, because it also functions as a measure of how two variables change together.

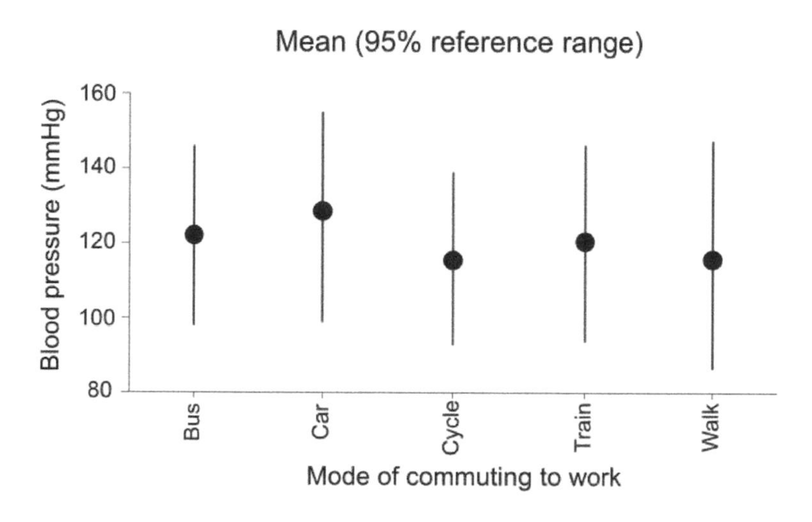

Figure 5.3 A dot plot with error bars.

5.3 QUANTILES AND OTHER ROBUST STATISTICS

The median has a more immediate interpretation than the mean. Half of the data lie on the median or above it, the other half on the median or below it. It is also more "robust": if there are errors in data collection which show up as very low or high outliers, these will pull the mean up and down. However, the median simply has to have half the data above and half below. It doesn't matter how far above and below, so errors at the extremes have no influence on the median, as long as the value isn't on completely the wrong side.

In the same way the median is a robust alternative to the mean, there are other statistics which can be used to describe the spread of the data. The most common are the quartiles, which divide the data into four equally populated parts. One quarter of the data should lie below (or on) the first quartile, and one quarter above (or on) the third. The second quartile is the same thing as the median. The first quartile is also known as the 25th centile (imagine dividing the data into 100 equal parts), the median is the

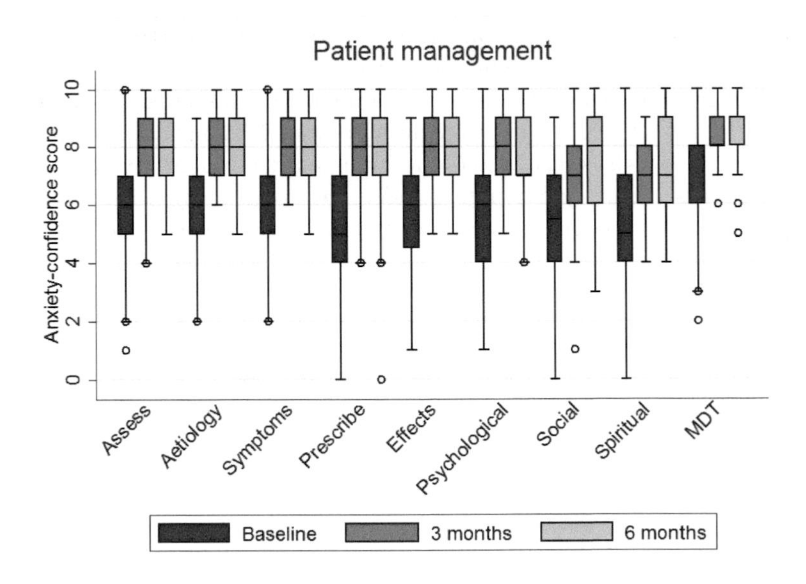

Figure 5.4 A boxplot comparing nine variables over three timepoints, created for Princess Alice Hospice, Esher, England.

50th centile, and the third quartile is the 75th centile. You can also work with different measures of spread, perhaps the 2.5th and 97.5th centiles, which together correspond to the 95% reference range we saw before.

With skewed data, quantiles will reflect the skew, while adding standard deviations assumes symmetry in the distribution and can be misleading. For instance, the Atlanta commute times have mean 18 minutes and standard deviation 14 minutes. We've seen that they are skewed in distribution: there are a few very long commutes. The reference range based on standard deviations will extend into negative commute times, which is clearly wrong. Instead, using the quantiles will get a better estimate of spread: 95% of the data lie between 1 minute and 50 minutes.

We can use dot plots to put markers at the median and error bars extending to some quantiles, as long as we make it clear what they represent. A popular format, expanding on this idea is the box plot (Figure 5.4).

A box plot shows five statistics, all encoded to horizontal or vertical position. The median is the line in the middle of the box, and the edges of the box are the first and third quartiles. There are then "whiskers" extending to the extremes of the data. But sometimes, outliers could be hiding under the whisker, so typically they will extend to the extreme data point or 1.5 lengths of the box (the box length is also called the inter-quartile range), whichever comes first. This convention is sadly not adopted by all software packages, so you may sometimes find that the whiskers do something different, and it is worth checking the documentation to be sure.

We can obtain a two-dimension variant on the quantile concept too. Imagine a scatter plot of points. If you could put an elastic band around the points, you can identify the outermost points as those that the elastic band touches. This shape that encloses the data is called a convex hull. If you remove them and repeat until you have removed, say, 10% of the data, then you will have a 90% quantile in a two-dimensional sense.

Quantiles are not the only robust statistics in town. Another popular approach to reducing the influence of outliers is to trim or Winsorize the data. Trimming involves simply setting aside any data outside a certain quantile before calculating the mean or other statistics. Leaving out the top 10% and bottom 10% would yield an 80% trimmed mean, for example.

Winsorizing is similar but replaces those outer data values with the quantile chosen as a threshold, so effectively they are replaced with the quantile, and then the statistics are calculated. It's worth thinking for a moment about why the 80% trimmed or Winsorized median would be no different than the original median. For the same reason, the inter-quartile range will not change until you trim or Winsorize 50% or more of the data, which would be an extreme intervention in the data!

For describing spread, you might want something between the standard deviation and the inter-quartile range for robustness, and the median absolute deviation could help. We calculate the mean, and then each value's distance above or below it (the deviation), we change the negative deviations so they are all positive, and then find the median of those. This is a nice way of describing scatter around the (possibly trimmed or Winsorized) mean because it is quite tangible: half the data are farther away than this distance.

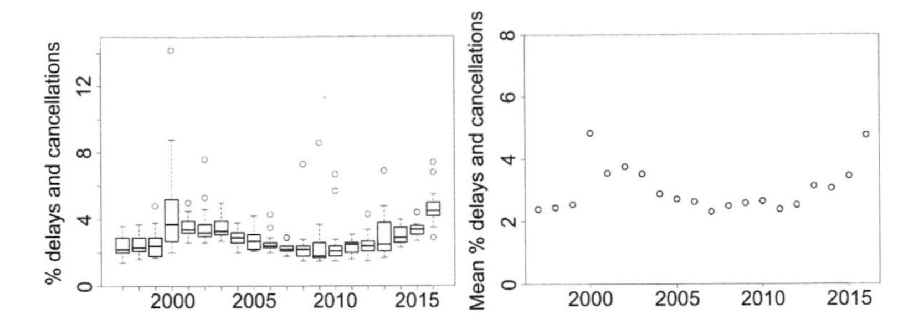

Figure 5.5 A box plot (left) and dot plot (right) summarizing change in train delays over 20 years. Notice how the mean for 2000 is inflated by just one exceptionally bad period, which is visible as an outlier in the box plot.

5.4 SMOOTHING

So far, we have been boiling our data down to just a few statistics. Sometimes we want more than that. Consider the train delay data from Chapter 2 again. The line showing time trends across all data points (Figure 2.2) was not helpful because it actually had too much data in it. To see the overall trends and patterns, we need to reduce our data down a bit more. This is perhaps the one thing statistics is about, more than anything else, and we'll look at it in more detail in Chapters 9 and 10.

To visualize the longer-term trends, we could chop up time into years and then draw some statistics for each year. Figure 5.5 has a box plot and a dot plot, while Figure 5.6 joins together statistics over time: quartiles and 80% Winsorized mean $\pm$ the median absolute deviation. Because there are 13 observations in each year, the 80% Winsorization affects the two worst periods and the two best periods.

None of these options is right or wrong – the choice should be led by helping the readers to understand the message most easily. There are always plenty of options for summarizing data and visualizing the summaries; never feel restricted to one choice that is popular or you have seen used before. You can do user-testing with different options to see which is understood most easily.

There are several techniques to smooth out these lines. It's easier for the reader to absorb smoothed lines, because there's less information shown, especially when the visualization gets cluttered.

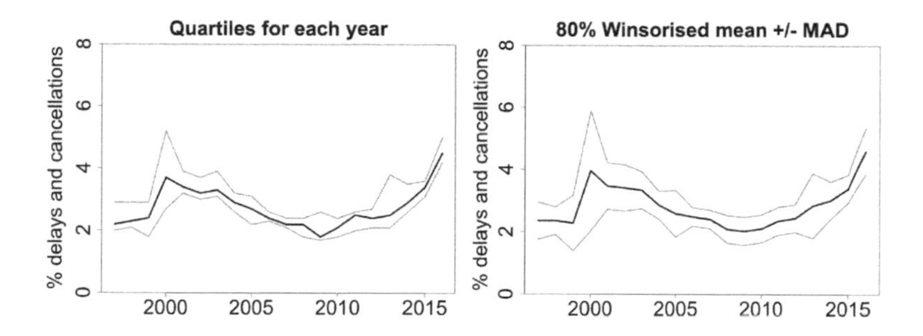

Figure 5.6 Lines connecting statistics to summarize change in train delays over 20 years: the quartiles (left) and 80% Winsorized mean ± the median absolute deviation (right).

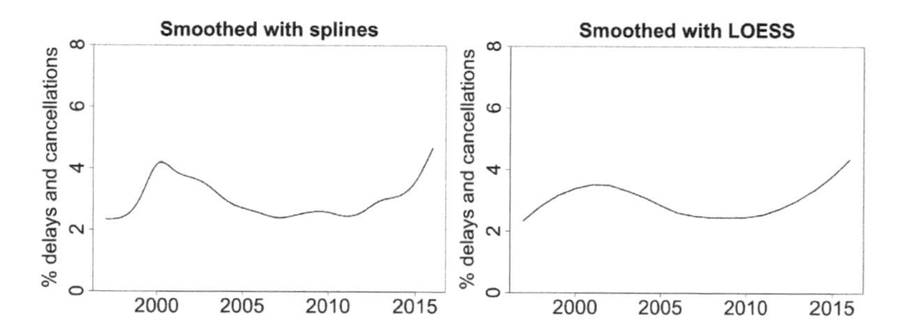

Figure 5.7 Smoothed data to summarize change in train delays over 20 years, using splines (left) and LOESS (right).

We've already seen kernel density plots doing this compared to histograms (Chapter 3). Without going into details, the most popular methods for putting smooth lines through scatter plots are called splines, LOESS and polished medians. Data analysis software can do these for you.

The person making the visualization can decide how much smoothing to apply: too much and the crucial details could be lost, too little and the line remains bumpy and might distract readers from the important trend. We will look at smoothing again in Chapter 10.

CHAPTER 6

Differences, ratios, correlations

OFTEN WE HAVE TO GO ONE STEP beyond presenting summary statistics: we have to present patterns in the statistics. Maybe we have a poll of voting preferences in various towns and we summarize those data as percentages. Our readers may also want to see what has changed since an earlier poll, or how the towns compare to one another.

6.1 DIFFERENCE OR RATIO?

Most of what we do to compare statistics falls into subtracting one from another or dividing one by another. Means and medians are well suited to subtraction: the difference in these statistics is roughly what the reader sees when considering adjacent violin plots, histograms, density plots, or whatever format. Their eye will generally go to the middle of the distributions and see how much it is shifted: a *change in the average*. If the data are matched and joined by lines, and it is not too cluttered, they will also get a general feel for the *average change*.

In the case of the mean for matched data, the difference in means is in fact equal to the mean of the differences. But this is not so for medians and other summary statistics, so visualizations of change in matched data should make it clear whether it shows statistics for the differences or differences in the statistics. If we have unmatched data, we cannot subtract or divide specific obser-

TABLE 6.1 Cancer diagnoses and person-years at risk (PYAR) in the Million Women Study, comparing number of weekly alcoholic drinks

Drinks per week	Million PYAR in the study	New cancer diagnoses
Non-drinkers	2.18	17,416
≤ 2	2.68	19,307
$3-6$	2.11	15,183
$7-14$	1.71	12,838
≥ 15	0.48	4031
All combined	9.16	68,775

vations by one another, so statistics of the differences are not an option, and can only really work with differences in the statistics.

Percentages, on the other hand, don't have such a visual clue for readers: there are just some bars in a bar chart, for example, and some are longer than others. We can subtract percentages to get a difference, or divide them to get a ratio.

Here are some figures analyzed by Dr. Naomi Allen and colleagues from the Million Women Study, which look at alcohol consumption and the appearance of cancer in the years that followed. Because not all women stayed in the study for the same time, a time at risk is calculated for each woman and these are added together to give the person-years at risk (PYAR) for the alcohol groups in question (Table 6.1).

We can divide the number of women with new cancer diagnoses by the PYAR and get an incidence rate. It makes sense to compare this rather than the number of women, because the low numbers in the group with heaviest drinkers indicate that there are not so many of them, rather than that it is safe to drink so much. This gives us rates in Table 6.2.

Clearly, the rates per 10,000 PYAR are easier to read and understand than the raw rates, and this kind of conversion to a more human scale is essential for good visualization. It would not be a good idea to show the rates per PYAR with an axis extending all the way to 100% (where everybody gets cancer within a year), because the values are so small that any markers or bars would just disappear, but also because 100% risk is just not of interest here. We can plot the region from 0 per PYAR to 0.01 per PYAR (0 to 100 per 10,000 PYAR), which captures all our stats (Figure 6.1).

TABLE 6.2 Incidence rates of cancer diagnoses in the Million Women Study, comparing number of weekly alcoholic drinks

Drinks per week	Rate per PYAR	Rate per 10,000 PYAR
Non-drinkers	0.0080	80
$\leq$ **2**	0.0072	72
3 – 6	0.0072	72
7 – 14	0.0075	75
$\geq$ **15**	0.0084	84
All combined	0.0075	75

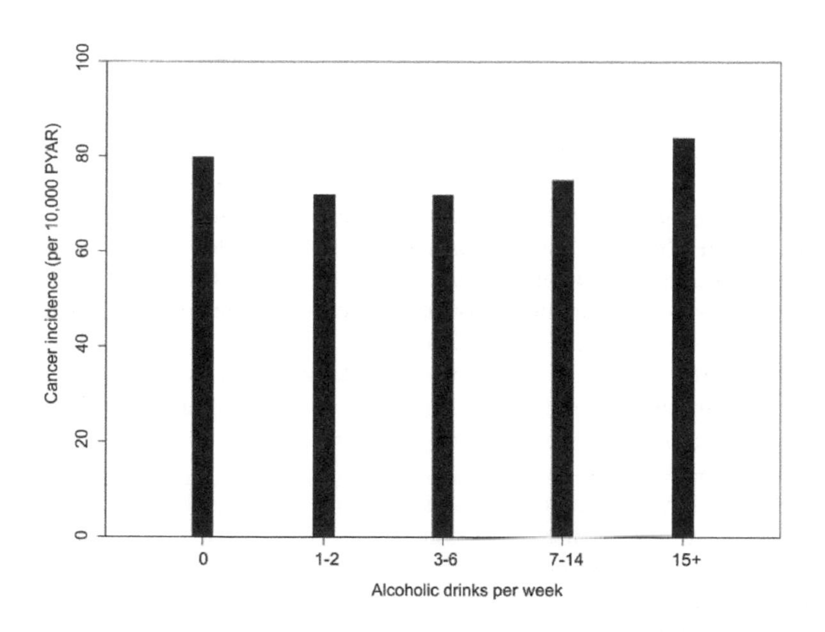

Figure 6.1 Incidence rates in the Million Women Study (unadjusted).

It might be helpful to contextualize these rates by annotating the bar chart with a horizontal line at the British incidence rate for women aged 55 (the average in the study), which is 56 per 10,000. This would show that the study has somewhat higher incidence of cancer than we would expect from the general population. Perhaps there is a bias leading women who know they are at risk to participate, or perhaps they are just diagnosed sooner when they are in the study. With the context made clear, the reader can think about this.

We could also add some unrelated incidence rates, like the regional variation. The Office of National Statistics gives incidence rates for men and women at all ages in England: from 57 in London to 65 in the North East. Adding this to the plot would help the reader see that the alcohol-related variation is greater than the regional variation. However, we have to be a little careful; alcohol consumption is likely to vary by region too, in which case maybe we are just showing the same thing two ways. If you are reading visualizations like this and they lack context, you should seek out some related statistics like I just did, and see where they lie on the chart.

By rounding off the rates per 10,000, we can make this a little more human again: an extra 12 women with cancer! These are called natural frequencies: the number of women expected on average. We might make it even more human with a pictogram, or some interpretive heading like "of 10,000 women who have one or two drinks a week, 72 will get a new diagnosis of cancer each year; this rises to 84 in women who have 15 or more drinks a week." You could even give a local example of a town with about 10,000 adult women, if you thought that would help.

But now we need to make the comparison, and we'll find that the human element is hard to maintain. We can compare each of the categories to those who have 1 or 2 alcoholic drinks a week (Table 6.3).

If we divide them to get a ratio, it gives a multiplicative increase, again assuming cause and effect. With ratios, a value of one means no difference (anything multiplied by one is unchanged). The Million Women data gives rate ratios because we have divided our number of diagnoses by the PYAR and are dealing with rates. We could say – and visualize – that there is a 16.6% increase in cancers among the heavy drinkers compared to the lightest drinkers.

That sounds quite alarming, but that is because the rate is not much bigger after multiplication. Indeed, some people might mis-

TABLE 6.3 Additional cancer diagnoses per 10,000 PYAR

Drinks per week	Rate per 10,000 PYAR	Additional diagnoses
Non-drinkers	80	8
≤ 2	72	0 (Baseline)
3 – 6	72	0
7 – 14	75	3
≥ 15	84	12

TABLE 6.4 Rate ratio of cancer diagnoses

Drinks per week	Rate per 10,000 PYAR	Rate ratio
Non-drinkers	80	1.109
≤ 2	72	1 (Baseline)
3 – 6	72	0.999
7 – 14	75	1.042
≥ 15	84	1.166

construe it as *adding* 16.6% to the incidence (which would take it from 0.8% to 17.4%!). It will also look alarming in isolation (Figure 6.2). It is a good idea to show both absolute and relative risks (or rates, in this case). If you can clearly use natural frequencies, it will probably avoid all this confusion. (Sometimes, you may read about relative risks and sometimes about risk ratios, especially in medicine. They mean exactly the same thing!)

When we are dealing with continuous variables, the same choice of multiplicative or additive scales applies, though it is not going to introduce as much confusion as with binary values. It is quite common to see the baseline value shown as a value of 100, and any change is then relative to that: 120 means a 20% increase, and 75 means a 25% decrease. This is a multiplicative scale because you multiply the baseline value by the change. Alternatively, the baseline could be shown as zero and any change from that in positive or negative numbers: an additive scale.

One advantage of setting the baseline to 100 and showing multiplicative change is that different variables can be brought together despite having different values; perhaps we want to see hospital waiting times and teacher-to-pupil ratios in schools alongside our train delay data, in one visualization that considers investment in

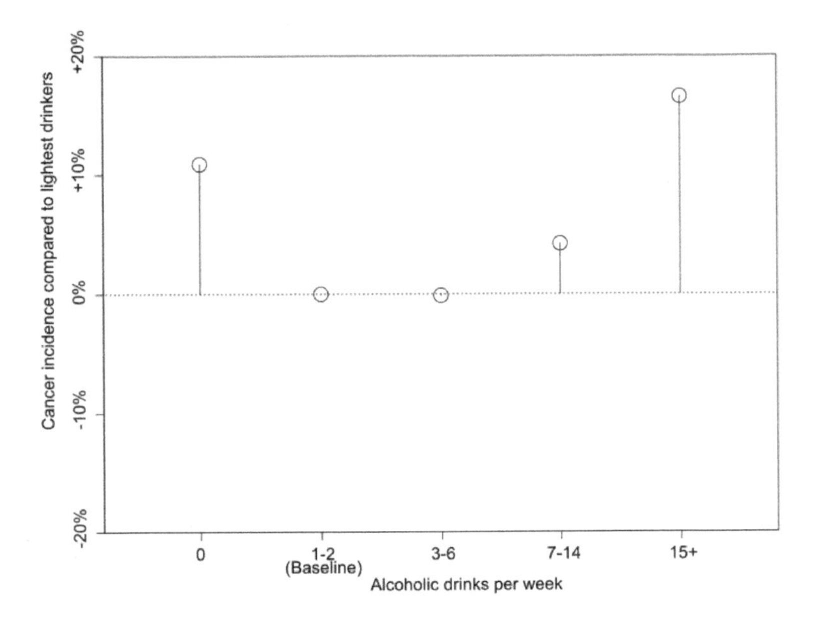

Figure 6.2 Rate ratios in the Million Women Study (unadjusted, and relative effects without absolute values). This format, essentially a dot plot with added vertical lines from the baseline, is sometimes called a lollipop chart. Leaving as much space for negative values as positive emphasizes the fact that all categories are at higher risk than the baseline, at the cost of some empty (but perhaps not wasted) space.

public services over time. In (Figure 6.3), we see two variables, wages and productivity, over time. They have very different values naturally but here we just see change since 1999 on the multiplicative scale and we can appreciate that wages have not kept up with productivity.

6.2 ODDS AND THE ODDS RATIO

Sometimes for technical reasons we have no choice but to work with the odds and odds ratio (which we encountered in Chapter 4). The odds ratio is always farther away from 1 than the corresponding risk ratio, so comparisons look more striking. Unfortu-

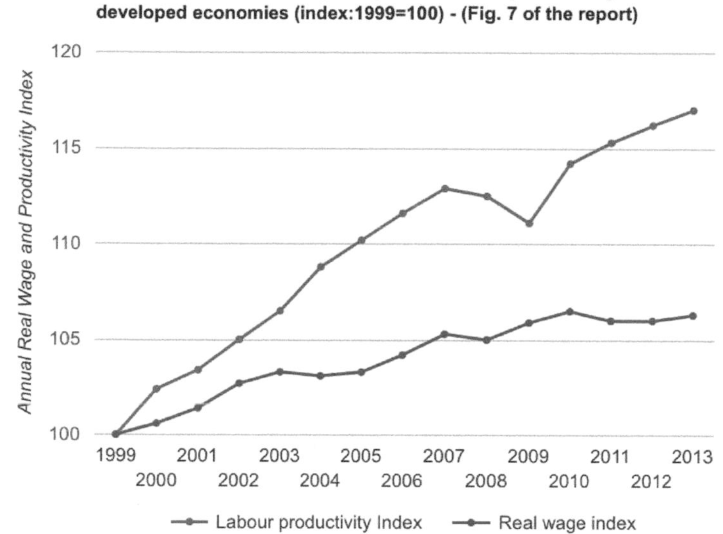

Figure 6.3 A line chart showing changes in productivity and wages since a baseline at 1999, which is given the index value of 100. Although the variables have different values, they are brought together into a common scale of relative change. Data source: International Labour Organization, reproduced with permission.

nately, researchers have repeatedly found that people (even those with scientific training) misunderstand the odds ratio and treat it as if it was a risk ratio. If you could simply convert the odds ratio to the risk ratio, this problem would disappear, but it isn't that simple.

In Chapter 10, I'll look at a study of different residential courses for drug addiction. For now, imagine you have the results of that study, and because some more complex statistical analysis was needed, this involves an odds ratio of 1.2, and we can interpret it like this: comparing two people who are identical in other respects but one went through the shorter course and one through the longer course, the odds of staying drug-free are 1.2 times higher in the person on the longer course.

So far so good; no matter what someone's odds are (determined in part by other factors like history of failed treatments), they will be better on the longer course. We can also say that the change in the odds is multiplicative: 1.2 times better for everyone. However, this doesn't mean the *risk* is 1.2 times better for everyone. The risk ratio will vary somewhat depending on those other factors. In short, an odds ratio shared by everyone doesn't mean a shared risk ratio.

What can we do? My advice for visualizing odds ratios is to convert them back to risks as much as possible. The risk is just a percentage or proportion, so it is easy to visualize and to interpret. Advanced statistical software can produce estimates of the risks (called marginal effects) over a range of different people, and make some nice charts of it too. We'll look at those in Chapter 10 where we dig deeper into the business of statistical models and this drug addiction study in particular. If marginal effects are not an option, you can still convert the odds ratio to a range of plausible risks and then show those.

6.3 CHOOSING A BASELINE

Calculating a meaningful difference or ratio depends on having a meaningful baseline. In Figure 6.3, we compared countries to their 1999 values, so those make up the baseline. When you are comparing groups, like taking a shorter or longer course to beat an addiction, one has to be the baseline. To simply say that the odds ratio is 1.2 is not enough; the reader needs to know which one is being compared. You could use the shorter course as the baseline, in which case the longer course has an odds ratio of 1.2 (improved odds of staying drug-free compared to the baseline). Or, you could regard the longer course as the baseline, in which case the shorter course has an odds ratio of 0.83 ($= 1/1.2 = 0.83$).

There is no rule to guide your choice of baseline, which you should choose to help the reader follow the story or message you want to communicate. To show how great the longer course is, use the shorter one as the baseline (odds ratio $= 1.2$). To show how the shorter course has been failing some of its participants, use the longer one as the baseline. If you have multiple visualizations tackling the same comparison, be consistent in your choice of baseline or you will confuse readers.

6.4 CHERRY-PICKING

But here is a warning. Although we can and should choose statistics to clarify our messages, we should not cherry-pick which parts of the data to show, and hide the rest.

When the spin-off analysis from the Million Women study I described earlier was published in the prestigious *Journal of the National Cancer Institutes*, there was a chart showing a relative risk arising from further statistical modeling, for various levels of alcohol intake (Figure 6.4). This roughly corresponds to the figures above and shows a steadily rising risk. The straight line looks quite compelling.

The journal asked an expert to write an editorial comment on it and they wrote "the message of this report could not be clearer. There is no level of alcohol consumption that can be considered safe." Other campaigners found this quite exciting too; the president of the Royal College of Physicians at the time wrote "we also have to be clear that there is no level of consumption that is risk-free, as clearly shown in the million women study."

They had all been tricked by the omission of the non-drinkers from the chart. To be fair, Dr. Allen and colleagues expressed concern about the non-drinkers – they felt that the data were misleading and the risk too high – and they noted in the paper that they could not distinguish people who never drank alcohol from those who only recently became non-drinkers.

This is a good point, but the same thing can be said of all the categories. It would have been better to include the data in the visualization, with an annotation about their concerns, and let the reader judge it for themselves. It goes to show that, if you are making visualizations, you should not put all the detail in a boring table and only the really exciting stuff in the visualization. Giving different layers of detail is good, but selecting what to show within one layer is potentially misleading.

If you are the reader, always consider whether some detail has been omitted by having a look at the tables, appendices and so on. It may be boring but you will be surprised at how often you are being shown a selected version of the facts.

6.5 CORRELATIONS

So far, we have been talking about differences across groups (unmatched data), or timepoints (matched data), or something sim-

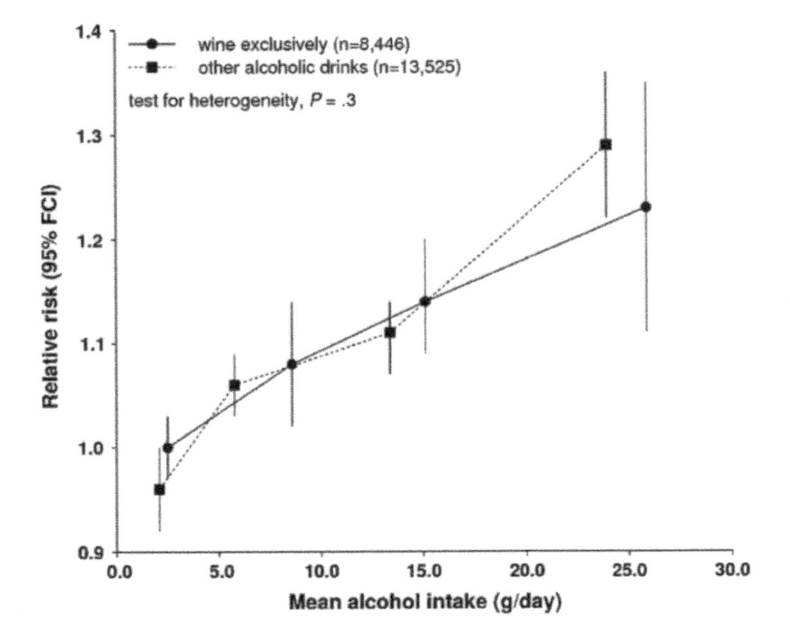

Figure 6.4 Line chart with error bars from Allen and colleagues' paper on the Million Women Study, reproduced with permission of Oxford University Press.

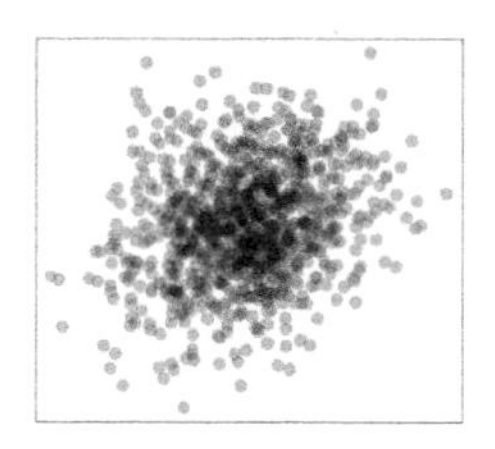
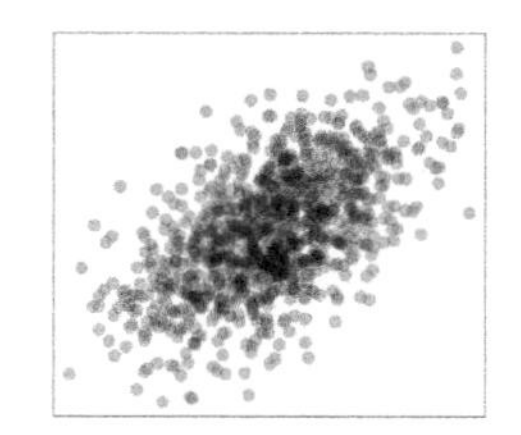
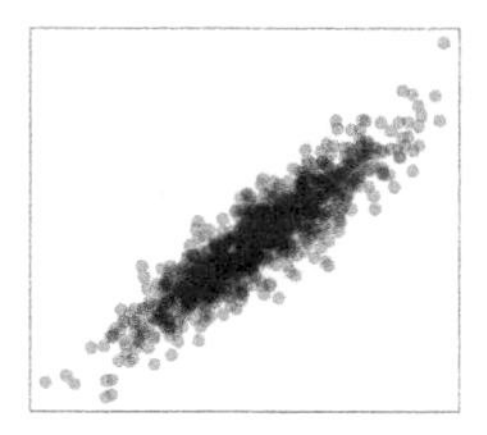

Figure 6.5 Scatter plots of 1000 random numbers with correlations 0.2, 0.6 and 0.9 from left to right.

ilar. What if we have two continuous variables and we want to describe how, as one goes up, the other tends to go up too (or maybe goes down). This is the role of correlations. Although that word is often abused to mean any generic association, it specifically means a statistic describing this numeric connection between two variables. It can take values from −1 to 1. Positive values mean that as one variable goes up, the other tends (on average) to go up too – as shown in Figure 6.5.

The more stretched out the markers are into a long straight line, the higher the correlation will be. It does not tell you how steep the line is, just how stretched out the markers are. Negative values mean that as one goes up, the other goes down, so the shapes in scatter plots will be stretched from top left to bottom right instead. A correlation of zero means a shapeless cloud of dots, neither sloping up nor down.

There are several correlation statistics, just as there are several ways of measuring how spread out data are. The most widely used is called Pearson's correlation, and if someone does not tell you which one they have used, it's probably Pearson's. A more robust version called Spearman's simply replaces the data values with their ranks (1, 2, 3, and so on from smallest number to largest number) and then puts those through Pearson's formula instead. The important difference this introduces is that Spearman's will detect any monotonic (always increasing or decreasing) relationship between the two variables, whereas Pearson's wants to find a straight line before it returns a high correlation statistic.

It is unusual to see correlations visualized like the dot plot, with just a marker for each correlation, encoded as the horizontal or vertical location, though it could be done. Shapes of variously

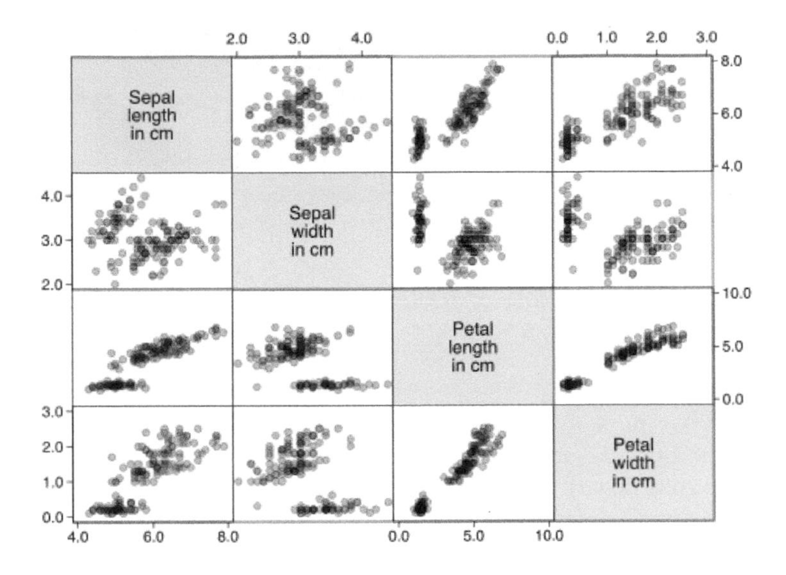

Figure 6.6 Scatter plot matrix of four measurements of iris flowers.

elongated ovals would be another obvious encoding, though it could only be appreciated at an ordinal level of information and would require a lot of statistical literacy. For most purposes, though, the scatter plot does the job adequately. Often, there are several variables and they all have to be shown correlated with each other.

The scatter plot matrix in Figure 6.6 shows this reasonably well. The markers have to be quite small and possibly semi-transparent, and the labeling of axes minimal, in order to keep it clear for readers. This is not a common visualization for lay readers, so some explanation is required. Personally, I like to introduce any matrix visualization like this by comparing them to the tables of distances that one sometimes finds with road maps, where towns are listed down the left and along the top of a table. If the reader has ever used them to look up the distance from A to B, they will know how to read these visualizations.

Look along the row labeled with a particular variable until you find it intersecting the column with the other variable you are interested in. That will locate a scatter plot showing those two variables together. The diagonal line down the middle is not needed

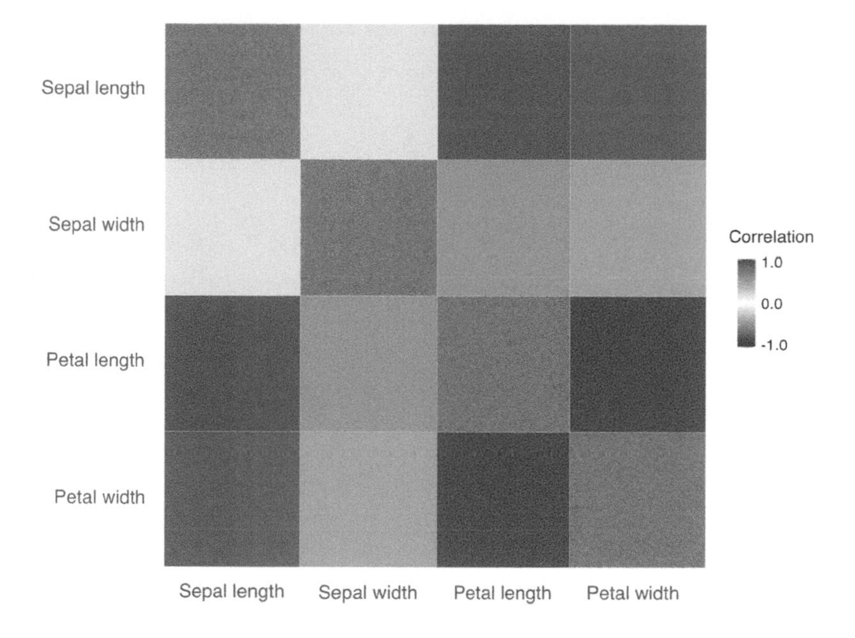

Figure 6.7 Heatmap matrix of correlations among four measurements of iris flowers. The color scale is diverging: it extends either side of a neutral value (0) to two different colors. All variables are strongly and positively correlated, except sepal width, which is weakly and negatively correlated to the others.

and in some versions simply shows each variable against itself as a straight line. Also, only one of the halves either side of that diagonal are needed, as they are mirror images of each other.

Another variant is achieved by dropping the scatter plots and just coloring in the blocks according to how high the correlation is. In fact, this is easily achieved in some spreadsheet packages once you have a table of the correlations, by using "conditional formatting." This approach of coloring in blocks according to some statistic that relates the horizontal position to the vertical position is called a heatmap (although the term heatmap gets used for other things too sometimes, like two-dimensional kernel density plots where density is encoded as color).

III

Specific tasks

CHAPTER 7

Visual perception and the brain

IT HELPS TO KNOW A LITTLE about how the human brain processes visual information. It's very popular to explain this in terms of evolution, even though it is largely speculative. Nevertheless, the great majority of our million years or so on earth involved finding things to eat and spotting predators before they spotted us. Sitting down and looking at data is a new preoccupation, but uses the same old hunter-gatherer apparatus (eyes and brain). We tend to notice only the very broadest outlines of our surroundings except for one or two things that stand out in some way and draw our attention.

As a first principle, any visualization should convey its information quickly and easily, and with minimal scope for misunderstanding. Unnecessary visual clutter makes more work for the reader's brain to do, slows down the understanding (at which point they may give up) and may even allow some incorrect interpretations to creep in. You might hear this called chartjunk. The designer Edward Tufte encourages us to think about the data:ink ratio, which you should try to keep high at all times. Statistician William Cleveland was more specific: the plot region is the part of any visualization that has to be clutter-free. That is the space between axes where the data appear. Annotations, examples, and even just eye candy outside the plot region impacts less on understanding. Sometimes a key, showing what different colors stand for, can be placed in the plot region without intruding much on the reader's attention.

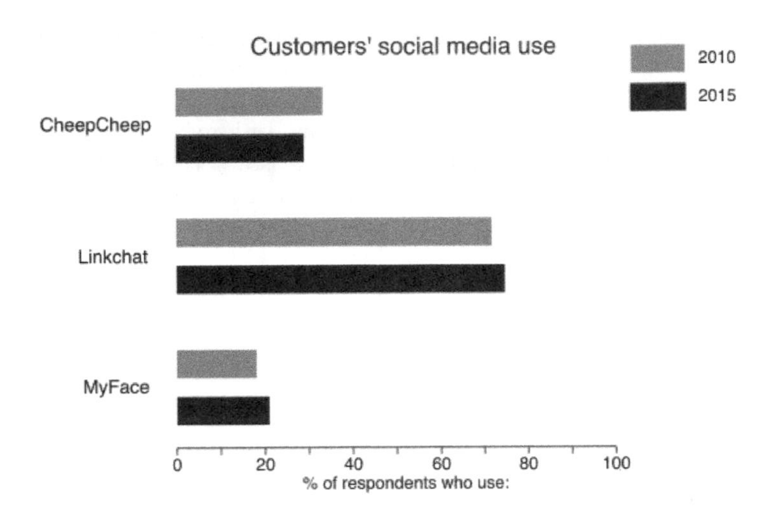

Figure 7.1 A version of Figure 4.6 with a better data-to-ink ratio.

Although simplicity helps to avoid distraction, and some designers claim that a good visualization will require no explanation, not even axis labels, a legend or key, most experienced data analysts recognize that some explanation and guidance is essential. Complicated visualizations can make use of a "How to read this chart" paragraph, and talking the reader through how to identify and interpret one of the aspects of the data can be helpful.

Getting the reader to understanding the visualization at the time is a different task than getting them to remember the image or its message. Some research has found that including relevant and witty chartjunk can actually help recall, but it has to be done carefully.

7.1 ATTENTION AND CLARITY

Often, a visualization tells a story or conveys one specific message out of a larger analysis. Scientific training discourages analysts from telling the reader what to think, but in dataviz it may be important. Such points of interest can be highlighted: a steady increase in a line chart or one out of a cloud of markers in a scatter plot, for example.

This can be done effectively without cluttering by using pre-attentive cues. These are features that our brains seem to be hard-wired to detect. We can add unobtrusive features to our visualizations just to help tell the story like this. In Figure 7.2, one marker is highlighted by color, another by size, then part of a line chart by shading around it. Very little is needed to draw the eye. I challenge you not to look at those points!

Crucially, these highlights have to be used sparingly. If there are too many of them, the reader will feel overloaded with information and they will no longer work. If you are making visualizations, be careful not to fall into a trap where you are very familiar with the data, so everything you create makes perfect sense to you. Experimentation and user testing will help you out. In Chapter 17, I am going to revisit some of these highlights and link them to everything else that surrounds the visualization.

We can also influence how the reader sees objects as being connected in some way. Good data visualization builds on the long-established Gestalt principles. The most obvious is that objects (like markers or lines) that are close together in a cluster and distinct from others farther away will be seen as connected. If we encode some of our variables as location or length then this follows naturally. But there are others that are not used so often:

- Draw subtle lines connecting the objects of interest together.
- Identify a group by a very distinct color and shape (for markers) or pattern and thickness (for lines).
- Enclose them in a shaded area, or surrounding oval or rectangle (more complex shapes will lose this effect).

Of course, it's not always possible to connect objects in the visualization without clutter, but it is worth considering. As with the pre-attentive cues, don't overload the reader. There should only be one group that gets connected per visualization for maximum impact, and going beyond this can backfire. If you have multiple messages, maybe you need multiple visualizations (or an interactive one).

Jittering takes objects that confusingly coincide on the visualization and moves them by small random amounts. Scatter plots with markers piled on top of one another now have a cloud of closely packed markers around a common point, and line charts with the same problem now have a bundle of lines moving closely together from one common end to another.

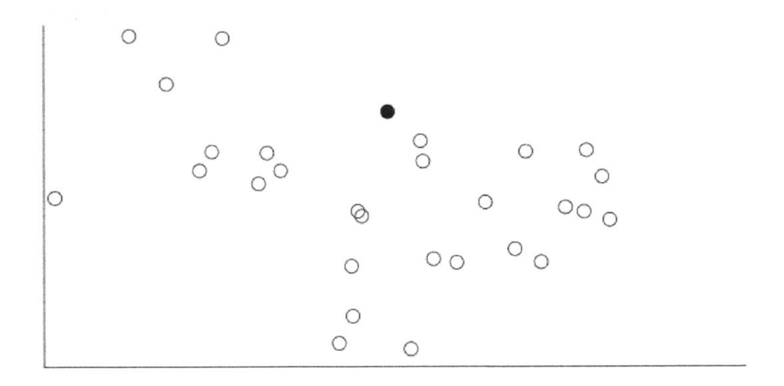

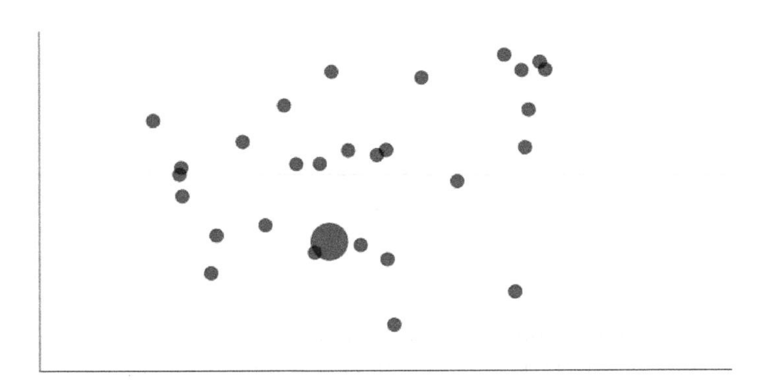

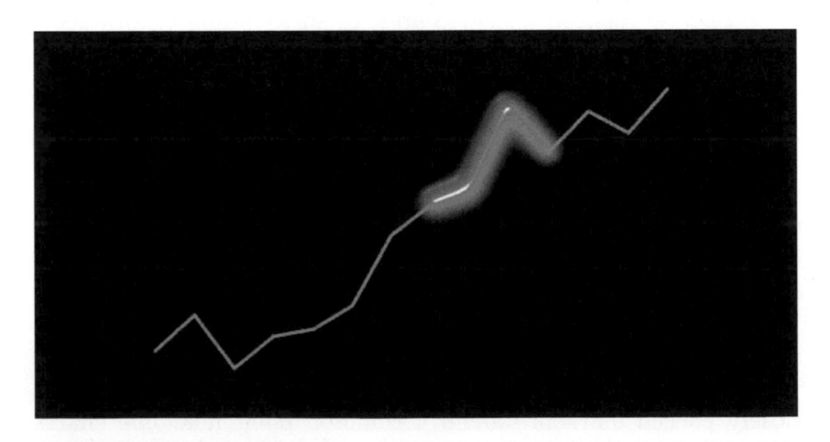

Figure 7.2 Examples of pre-attentive features.

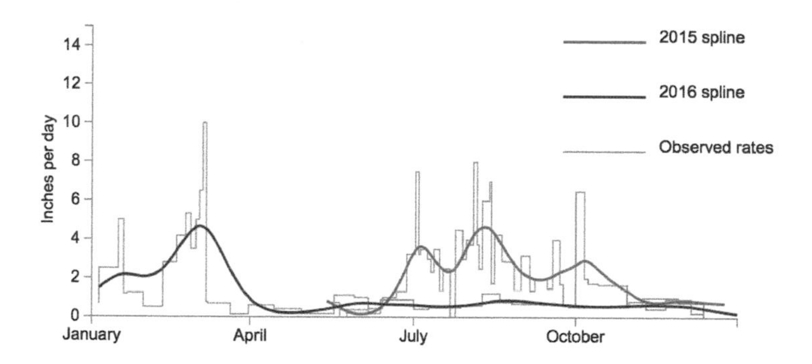

Figure 7.3 Observed rates of bird seed consumption in my garden, and smoothed lines through the data using splines. robertgrantstats.co.uk/dataviz/birdfeeders

Smoothing is perhaps the opposite of jittering, in that a lot of information gets summarized into one simple impression. A curve wiggles through a scatter plot, tracking the markers, or through a chart with multiple lines, showing a summary of the patterns (Figure 7.3).

The important feature of smoothed curves is that the smoothness is not part of the data. In the bird feeder data of Figure 7.3, the consumption often changes, and the resulting lines are very rough series of steps up and down (the gray lines). Most readers would find it hard to see the overall pattern, but the smoothed lines make it easier: more seeds consumed in summer 2015 and spring 2016, less so in the winter and through the rest of 2016. We are compromising by bending the natural line of the data, with the intention of improving understanding.

In Chapter 10, we'll explore different techniques for smoothing in the context of models that predict one variable based on others. If your aim is not as formal as all that, and you just want to give a simplified impression, you could try a trick suggested by John Tukey that didn't catch on: instead of small markers like circles, draw a vertical line for each data point. The overall shape will be apparent to readers but the central locations on the line will not be obvious (Figure 7.4).

Sometimes, there is a good reason for breaking a sequence of data into more than one smoothed line. For example, if you have economic data before and after the credit crunch of 2008, then

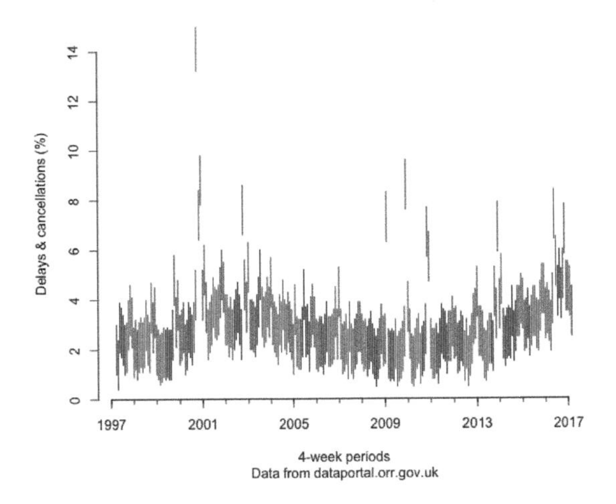

Figure 7.4 Tukey's smoothing by drawing vertical lines instead of points, applied to the train delay data from Chapter 2.

you know from the context, even before you draw the data, that it could be represented as one smooth curve before and another smooth curve after the crunch.

Another simplifying trick which we will encounter in Chapter 11 is edge bundling, where lines connecting points together are artificially pulled together to reduce the spaghetti effect.

Semi-transparency is a great all-round tool for busy visualizations, also known as opacity. This allows lines, markers and such that coincide to be seen. Those in the background show through slightly. When markers are piled on top of one another, they look extra dark compared to others on their own. Because semi-transparency is more like the real world, we get an impression of lines moving continuously over and under one another and are able to take in more information immediately. There are several images in this book with semi-transparency, such as Figure 8.2; even though there are many markers or lines, you can see where they pile up in greater numbers.

Colors, lengths, and areas are some of the attributes to which we have been encoding data. These are stimuli that get perceived by the brain. Not all stimuli have the same effect; the psychologist Stanley Smith Stevens showed that, if you double a length, it will be perceived accurately as twice the size of the original, but doubling an area is underestimated as 1.6 times bigger, while doubling the redness of a color is overestimated as 3.2 times bigger. This is why serious data visualization experts don't like encoding things to area or color unless they are just ordinal (or you are happy for them to be understood as such).

7.2 CULTURAL ASSUMPTIONS

In many of the visualizations we've seen so far, time has been encoded to the horizontal location, with old data on the left and new data on the right. Why? This is an artifact of reading from left to right, and is so universal in dataviz that it is preferred even by writers of right-to-left alphabets like Arabic. Colors, too, do not have a universal meaning. Red is dangerous in some places and auspicious in others. It is a good idea not to assume your reader understands this sort of culture-specific encoding.

Some visualization formats are themselves cues to interpret the data in a specific way. For example, connections between data points have been visualized in the style of a subway map, and lists of items in the style of a periodic table of chemical elements. The trouble here is that, unless these are aimed wholly at city dwellers or chemists, not everyone will know what you are implying by the format. Although they are creative and fun, creators of these sorts of visualizations have sometimes been mocked for not having understood the thing they imitated. Do the distances between the subway stops represent anything? Is there actually periodicity in what looks like a periodic table, or is it just a glorified list?

7.3 LEARNING FROM OPTICAL ILLUSIONS

In data visualization, optical illusions are not just fun but actually give us some clues as to ways that people might misinterpret our work.

The café wall illusion (Figure 7.5, left) is one that may well affect data visualizations with blocks of color, causing lines to appear sloped when they are actually not. Lines entering and leaving shaded regions can also appear to bend (Figure 7.7), and wavy lines appear flatter or taller than they really are (Figure 7.5, right). Any visualization with high-contrast blocks of background color might be at risk from these effects.

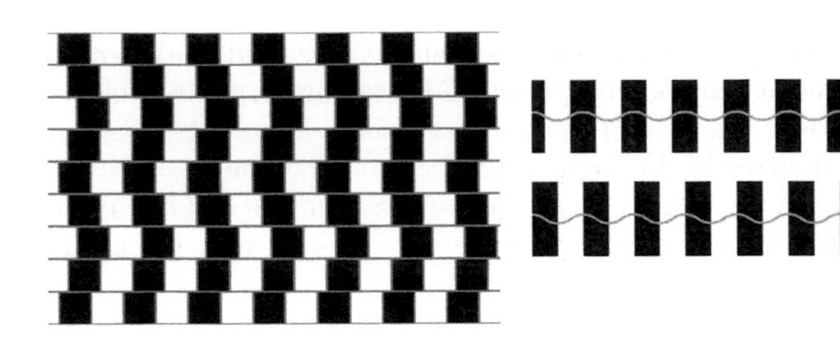

Figure 7.5 The café wall illusion (left), where all lines are actually straight and either vertical or horizontal. A related illusion by Akiyoshi Kitaoka (right), where the two gray waves are identical in height. Left image by Wikipedia user "Fibonacci" - Own work, CC BY-SA 3.0, https://commons.wikimedia.org/w/index.php?curid=1788689. Right image by Akiyoshi Kitaoka, used with permission.

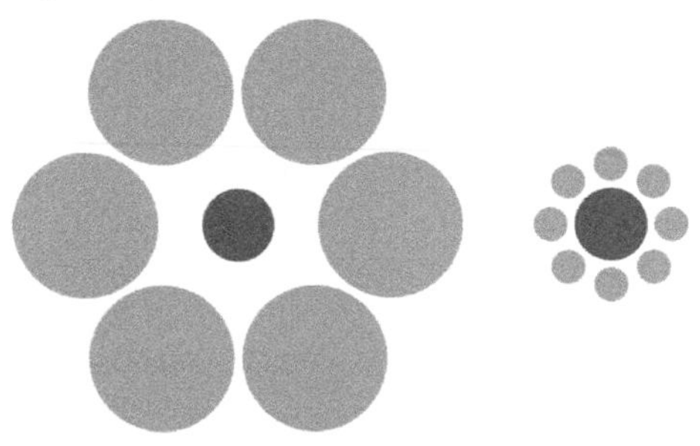

Figure 7.6 The Ebbinghaus illusion: the darker gray circles are the same size.

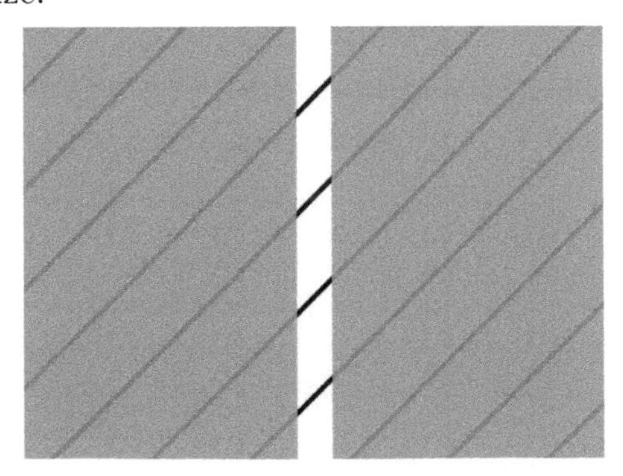

Figure 7.7 An illusion by Akiyoshi Kitaoka: the diagonal lines are all straight and parallel, used with permission.

The Ebbinghaus illusion shows that the perceived size of an object is influenced by nearby objects. Bubble plots could be affected by this, where the size of markers represents a variable. Our impression of colors is also influenced by the colors of nearby objects, and by what we expect to see. Any visualization with variation in the size or color of markers which are close together could be affected.

CHAPTER 8

Showing uncertainty

ESTIMATES BASED ON DATA are often uncertain. If the data were intended to tell us something about a wider population (like a poll of voting intentions before an election), or about the future, then we need to acknowledge that uncertainty. This is a double challenge for data visualization: it has to be calculated in some meaningful way and then shown on top of the data or statistics without making it all too cluttered.

The most common source of uncertainty is when our data are a sample from a wider population and we want to use them to tell us something about the population. Because our estimates are based on a sample and not the whole population, they give us an underestimate or overestimate, and we don't know which, but we can at least put some boundaries on that error.

Other sources of uncertainty include missing data, bias in data collection, imperfect measurements, and so on. We can attempt to quantify these uncertainties too, and if we do so we can visualize their impact, but there will be assumptions involved and these need to be made clear to the reader.

8.1 THE BOOTSTRAP

Ideally, we would run our data collection again and again, and see how different the estimates were each time. If they were really unstable, maybe we should not draw any conclusions. If they are always about the same, we can be more confident. Unfortunately, re-running the collection many times is not an option, but we can do the next best thing: we can make new versions of the data that behave as though they came from re-runs.

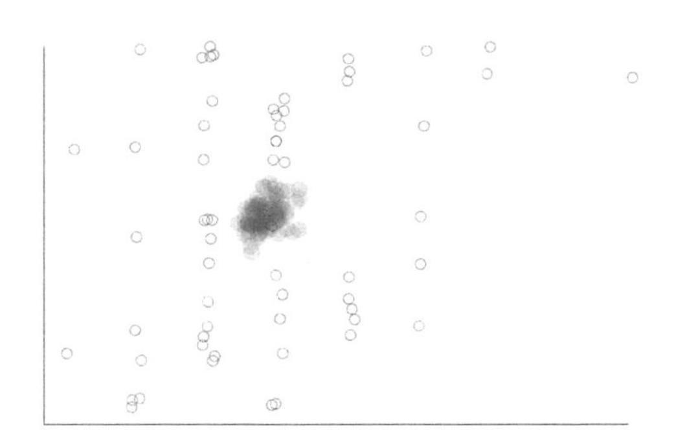

Figure 8.1 Semitransparent markers for 100 bootstrap estimates of the mean in two-dimensional data.

If we then calculate our statistics or whatever we need from each of those new versions and look at the distribution, it turns out we can get an accurate measure of the size of the uncertainty. This technique, which may sound too good to be true at first, is called the bootstrap.

If we are showing the mean, for example, and we generate a thousand bootstrap versions of the data, we will have a thousand versions of the mean. These scatter about the observed mean that we originally got (Figure 8.1), and if we discard the top 25 and the bottom 25, we will have a 95% confidence interval around our estimate.

In other words, there's a 95% chance the population mean is in that interval. (Some people prefer a more long-winded definition to do with infinite repetitions of the data collection and analysis, but I believe these come to the same thing. The difference of opinion is down to the philosophical definition of probability. I only mention it here so that, should you one day explain it in these terms, you can at least be prepared for the pedantry that follows.)

We can show this interval using error bars, as we saw for the standard deviation in Chapter 3, but now you can see why it's essential to explain what the error bars stand for: standard deviation, confidence interval, or another statistic called the standard error.

With some relatively simple calculations, like the mean, there is no need to bootstrap, because a shortcut formula can give you the width of the confidence interval straight away. This is generally done in terms of the standard error. Just as two standard deviations show the 95% reference range (within which we expect to find the

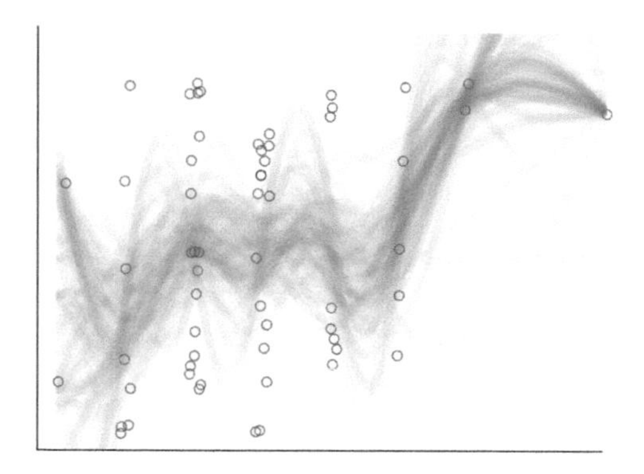

Figure 8.2 Semitransparent lines for 100 bootstrapped splines through two-dimensional data.

population data), two standard errors show the 95% confidence interval (within which we expect to find the population mean).

Another advantage of the bootstrap is that it cannot extend into impossible values, such as less than 0%. Some of the shortcut formulas can do this, which may lead readers to question the reliability of the whole analysis.

If the data includes the whole population, meaning there are no other observations to infer to, then there is no need for confidence intervals or other measures of sampling error. Inference to the future is different, though, and we return to that in Chapter 9.

8.2 CONFIDENCE REGIONS

Apart from error bars, there are a few options to show the uncertainty:

- Shade the areas above and below lines, or around markers, to indicate the confidence interval.
- Draw faint semi-transparent markers drawn at random from the bootstrap statistics (Figure 8.1).
- Add lines or shading for a series of quantiles from the bootstrap statistics, or theoretically from the standard error (like Figure 5.6).
- Draw multiple semi-transparent lines (Figure 8.2).

- Thicken, or make a line more transparent as it becomes more uncertain.
- Use interactive graphics (Chapter 15) or animation (Chapter 9).

The funnel plot is a compact alternative, if you want to compare one statistic across lots of groups of data. One example of this might be to look at hospitals' death rates in treating some life-threatening condition called Disease X. Although we might prefer to go to the hospital with the lowest death rate, if that is based on only a few patients, then it carries a lot of uncertainty with it. In a funnel plot (Figure 8.3), the individual hospitals (or other units to be compared) are represented by markers, the percentages (or whatever statistic) are encoded to the vertical position, and the number of patients (or other units making up the data) to the horizontal position.

Then, the hospitals with few patients are on the left and those with many patients are on the right. We would expect the hospitals on the left to be more variable just by chance, so we can superimpose a funnel shape that represents the confidence interval around the average percentage. Any marker outside the funnel indicates a clinic whose difference from the rest is so large that it is unlikely to be attributable to a run of bad or good luck. This sort of plot is routinely used to detect surgeons whose practice is potentially dangerous and warrants investigation.

Two-dimensional confidence intervals are sometimes seen by crossed error bars or shaded ellipses around markers. Unfortunately, the plot region will quickly get cluttered with these. It is simple to calculate the confidence interval for two statistics separately (one on the horizontal location, the other on the vertical), but bear in mind that they can also be correlated, so that a confidence ellipse might be tilted. Bootstrapping is a simple way to assess this.

One thing that definitely should not be done with error bars is to have them in a bar chart, appearing above the bars but obscured below. This seems to be standard practice in some scientific fields. Hiding some of the statistical information for the sake of a format is never a good idea. A dot plot is preferable if you need to show a few values with error bars.

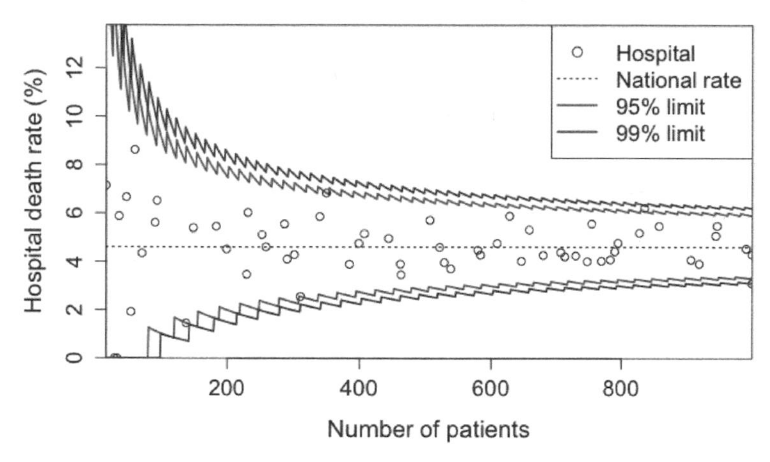

Figure 8.3 A funnel plot comparing fictitious death rates across 60 hospitals. Note that the confidence intervals are jagged because they show the *number* of deaths, which are fairly small and go up or down in steps. Often, alternative calculations are used to obtain smooth curves. The two hospitals at the bottom left with no deaths only had 27 and 31 patients with Disease X, while others were treating hundreds of people in the same period. Because the probability of lying outside the 95% interval is one in 20, and we comparing 60 hospitals, we would expect there to be a few hospitals outside the 95% limit just by bad or good luck, even if they were no different from the national death rate.

8.3 OTHER SOURCES OF UNCERTAINTY

Imagine a pre-election opinion poll, where people are called at random from the phone book. Those with particular views might have refused to take part, skewing the results. Maybe some people are not in the phone book, or are out at the time of day when we called, or don't own a landline phone. How can we account for these biases at the point of data collection?

There are two broad approaches. One is to present the analysis of the data as it is, and list the potential faults and biases in accompanying text. Unfortunately, that may well be skipped by the reader. Another is to try to take account of the biases and what effect they might have had on our observations. That will give us a more accurate result – but only if we get our assumptions right.

If we use Bayesian statistics for the adjustment, we can apply probability to any unknown, such as the association between political views and being away from home at the time that the call was attempted. We can get the computer to simulate different scenarios like this and feed them through the analysis one at a time. This gives multiple estimates, just like we have from the bootstrap, and these can be visualized in the same way.

Missing data is a common problem. One Bayesian method for this that has become very popular in recent years is called multiple imputation. It fills in the missing values with multiple informed guesses. Just one guess might be good but would imply that there was no uncertainty about it, so it is better to have several guesses that reflect the uncertainty too.

We can show the multiple imputed values in a faint, semi-transparent color, or with smaller markers, dotted lines, or some other way of making them carry less visual weight.

Data which we know are heaped (where data get rounded off in some instances and not others – for example, the number of cigarettes people recall smoking per day) or coarsely measured (where you know it's in a certain range but not a precise value – for example, if a doctor records patients' blood pressure as low, normal or high without recording the actual values) can also be imputed in the same way.

Weather maps, particularly for hurricanes, have popularized multiple curves showing uncertainty in predictions, so that if you use one of these approaches, it is likely to be understood quickly by readers. These don't only work for geographical data (we'll look at maps in Chapter 13).

In Figure 8.4, there are three different approaches: multiple estimates with semi-transparency (left), an interval around a best estimate (center), and contours showing that the edges are less likely than the central path (right). The center image suffers from the misconception that if you live just outside the interval, you are completely safe, and if you are inside, the hurricane is certain to pass over you. In other words, presented with an interval for

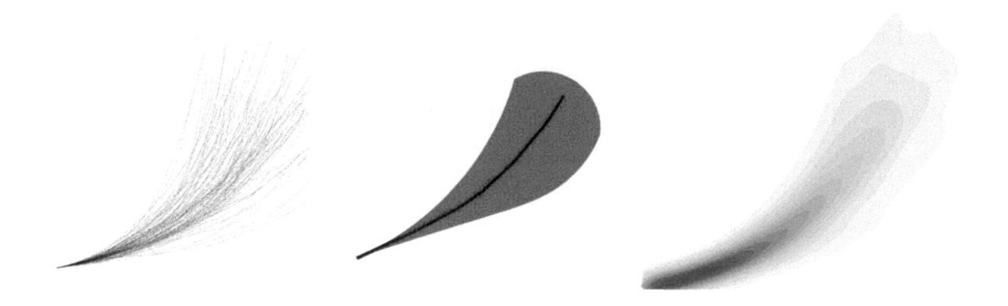

Figure 8.4 Three ways of showing uncertainty in projected hurricane paths – and other visualizations too.

uncertainty, people often interpret it as something else: statistical uncertainty is a difficult concept to grapple with. These visual approaches can all be adapted for lines and markers.

CHAPTER 9

Time trends

TRACKING CHANGES IN DATA OVER TIME is very common and we have already encountered some visualizations with time encoded to the horizontal location, using bars, lines and areas to indicate the data required. We call this time series data, and there are some special statistical techniques for analyzing it, which can impact on visualization.

9.1 MORE FORMATS AND ENCODINGS FOR TIME

I've already looked at bars, scatter plots and lines, which can all be used to encode a variable you are interested in to the vertical position and time to the horizontal. The areas under lines can be shaded in too. Bars and areas can be stacked, though this is hard for readers to digest; in recent years a lot of streamgraphs have appeared which have a smooth flow of brightly colored areas above and below a horizontal axis. They are striking but very hard to translate back into numbers.

The alternative to stacking is to have multiple lines or bars interwoven, one for each variable, group or observation, which can become hard to read as the numbers increase, and depending on how much variance there is within groups (jumping up and down and crossing each other). We can also choose to show the data or change relative to a baseline (see Chapter 6 for more on this).

A connected scatter plot shows how two linked variables have changed over time. They get encoded to the horizontal and vertical locations, and then the markers are joined by curving lines that trace out time (smoothers are useful here to help readers disentangle lines). So, time is not encoded as such but appears thanks

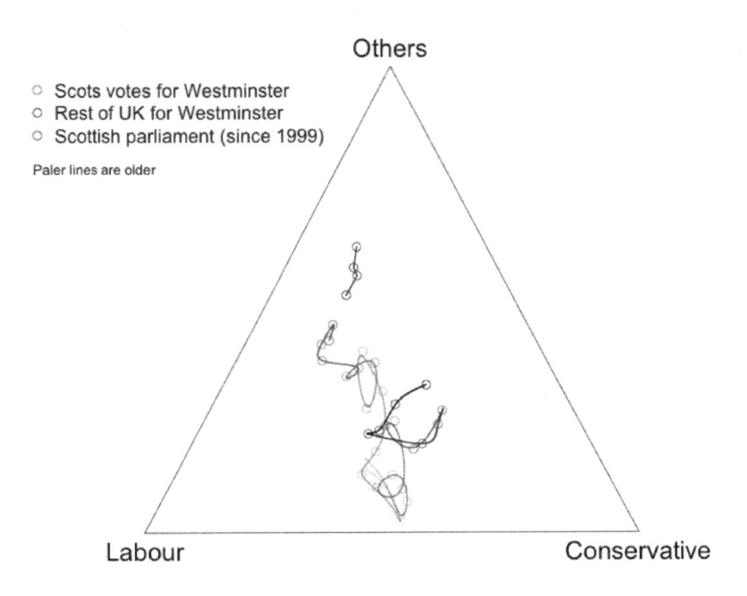

Figure 9.1 A connected scatter plot in a ternary layout: Scottish election results. This image uses splines in both horizontal and vertical dimensions to obtain the smooth curves like handwriting.

to the format. However, time could also be encoded in the color, size, or some other attribute of the markers and/or the line joining them (Figure 9.1). Animation can also capture time explicitly; the various talks of global health professor Hans Rosling, which can be viewed online, are excellent examples of this.

If we have a scatter plot showing two variables, and measurements were made at successive points in time, then the markers could be replaced with an arrow, tracing out the direction of change. This is very effective if the arrows do not overlap too much and if there is a consistent pattern. Chaotic arrows going in all directions are worse than no arrows at all, but some smoothing might be possible (Figure 9.2). Thin isosceles triangles or tadpole/comet shapes might be better than arrows, by doing away with the arrowhead that complicates and puts more "ink" onto the "page."

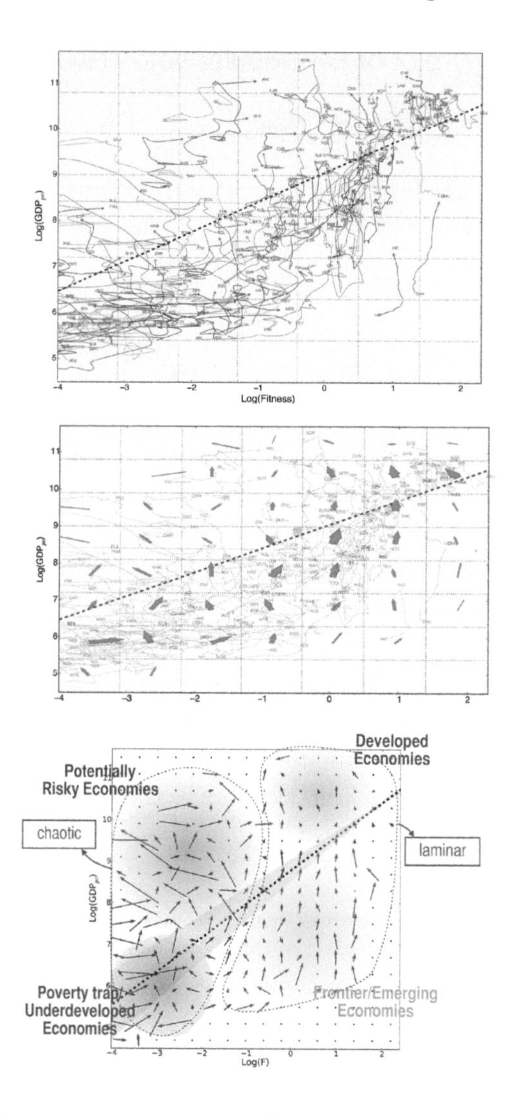

Figure 9.2 A connected scatter plot showing countries moving over time with Gross Domestic Product on the vertical axis and a measure of economic fitness on the horizontal axis. Irregular movement (top) is reduced to smooth flows by averaging countries' directions within each square of a grid (center). Interpretation is then based on the averaged arrows alone (bottom). From "The Heterogeneous Dynamics of Economic Complexity" by Matthieu Cristelli and colleagues, reproduced under PLoS One open access.

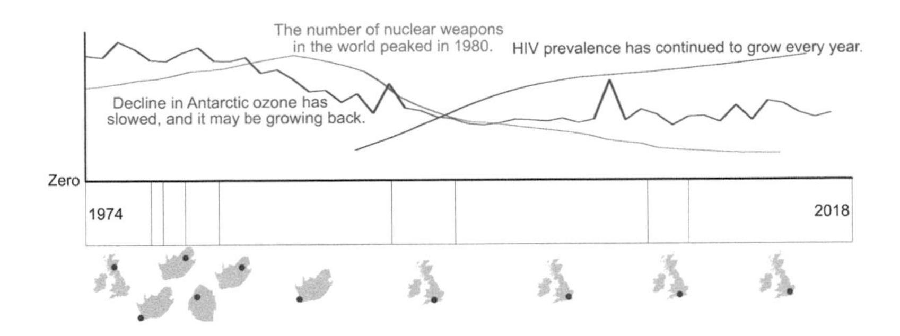

Figure 9.3 A timeline: all the cities I have lived in, plus mini maps, plus three line charts for things we lost sleep over in the 1980s. Some got better, some got worse. Note that the three line charts are all on different vertical axes, none of which are shown because I just wanted to convey the direction of change. This sort of broad-brush approach is likely to irritate more statistically literate readers.

Sparklines are very small line or bar charts for time series data, with no labeling, that can be incorporated into text like this: (that's the train delays from Chapter 2 in 2007, with a blue dot for the coldest period and a red for the falling leaves). They are useful for giving a quick impression of a pattern without taking up much space, and also have the benefit of breaking up text and being fun.

Readers will find it easier to see vertical rises and falls in areas, lines, or bars if the overall gradient (assuming there is a general upward or downward trend) is not too flat or steep – 45 degrees might be ideal – on the page or screen. You might need to stretch one of the axes in order to achieve that. William Cleveland suggested that long time series could appear in a sequence of vertically stacked charts to get closer to this 45-degree slope, which he called cut-and-stack. The disadvantage of this is that you break up the series, making it much harder for a reader to mentally join the parts back together again.

Timelines simply break up a horizontal or vertical length into chunks of time. Having done that, they provide a simple and informative framework to which we can add other charting details or small multiples (Figure 9.3).

In circular formats, some measure of time is encoded as an angle

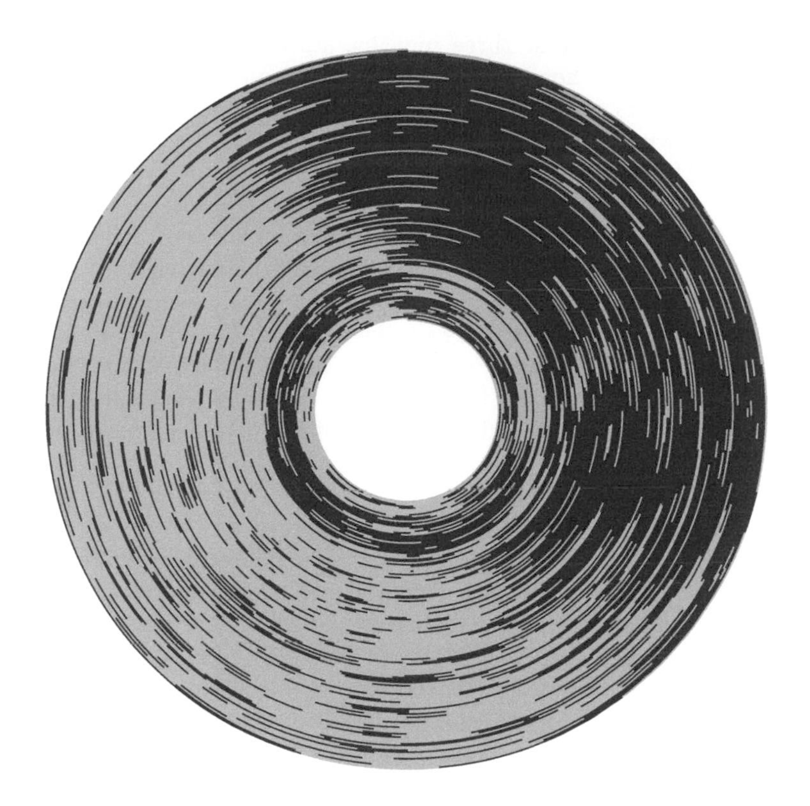

Figure 9.4 Waking and sleeping time over the first three months of a baby's life, in a spiral format. There's a painful (for the parents) flip in and out of being mostly nocturnal in the first few weeks. Copyright Andrew Elliott (andrewelliott.design), reproduced with permission.

around a central point, and a variable of interest is the distance from that point. This can be effective if there is a strong periodic pattern, but otherwise is often too confusing.

If we want to show hours within days, days within months, or similar, these are nested loops, and we could capitalize on this in the visualization, using some spiral format. To represent this in two dimensions requires some compromise. If lines spiral out from a central point, they will occupy more space in the image in the outer loops of the spiral than they do close to the center, but that might not affect understanding. Figure 9.4 shows a baby's sleep patterns over the first three months, starting at the center and moving out. The hour is encoded to the angle around the circle, and the day is encoded to distance from the center. Each day is a circuit, with waking time encoded as beige and sleeping time as purple. Midnight is at the top.

To provide an eye-catching visualization, or when all the data simply cannot be clearly shown in one chart, we can use animation with great effect to show the numbers changing over time. This effectively encodes time to time, though maybe speeded up or slowed down. Hans Rosling's talks with animated bubble charts were a landmark in data visualization in the 2000s, and they have been watched online millions of times. You can make animated charts as a stop-frame animation out of lots of static image files. This book's website links to some advice on this.

9.2 STATISTICAL CONSIDERATIONS

Regression to the mean is a common problem in data over time. Imagine we are working on traffic safety. The city has paid to install speed cameras at its most dangerous junctions, based on last year's data, and now we must show whether this has been effective. We compare this year to last, and find an improvement!

Was it the cameras? Perhaps not, because some years are just randomly bad years at some junctions, and in contrast next year will tend to look better, even if the cameras did nothing. That's to say, sometimes random noise happens in your time series data and if you select data on the basis of very high or low values, then you might find the noise tends to go in the opposite direction (Figure 9.5).

This is a concern for data visualization because we often select a part of the data on the basis of baseline values and visualize it. The best way to avoid this is to have plenty of data from all sorts of baseline values – but that's not always an option. As we found in Chapter 6, it's best to show both change relative to baseline and the absolute values. For changes over time, like the traffic cameras, it would be helpful to show a longer period of time. If the problem was just transient, this would stand out as a spike. We could also show the spread at the baseline (accident rates at all the junctions in the city, for example) and where the selected data lie in that.

Davis Balistracci has written an amusing, but worrying, list of the ways in which any three timepoints of data can always have a political spin to sound good (explained nicely on the Thinkpur-

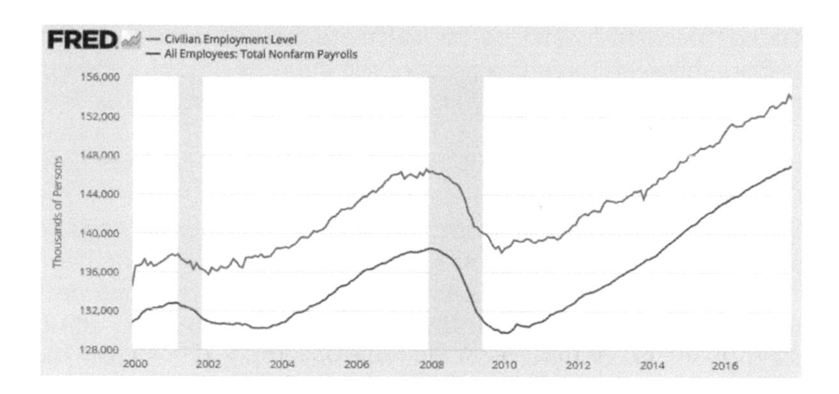

Figure 9.5 American unemployment rates, 2000-2017. If only the period since the 2008-9 "credit crunch" was examined, a misleading pattern would be found because of regression to the mean. At the start of 2014, despite employment still being below 2008 levels, an unscrupulous politician could present only data since the start of 2010 and claim to have created 6 million jobs in steady growth. Created by Wikimedia user "Farcaster," used under CC-BY-SA 4.0 license.

pose blog at goo.gl/ohEbQZ). When you are reading a visualization showing a trend over time, and the people who prepared it stand to benefit from a particular message, always ask yourself what time period it spans, and whether that is really justified. At the time of writing (2017), politicians still regularly take the credit for economic improvements since 2008. But 2008 saw a massive crash in the world economy, so it is hardly surprising that things might look better now.

A special use of correlation comes into play with time series data. We expect one value to be similar to the one that immediately preceded it, and we can calculate a correlation or similar statistic between these two parts of the data. In statistics, we say the previous data is lagged. This is called autocorrelation (the data is correlated with itself), and it can show us some patterns that might not be so easy to pick out. In the train delays data from Chapter 2, we can spot the roughly annual cycles of delays by finding autocorrelations at one month's lag, two months', and so on, and then drawing these as a bar chart.

Another important problem to be aware of is changing definitions in second-hand data. If the data were not collected afresh for the purposes of the current analysis and visualization – perhaps they were extracted from an existing database with other everyday uses – then sharp changes may indicate that something has changed in the way that the data are collected, labeled or calculated. If they are drawn from something like a database of traffic accidents, for example, we might find that at one point the law changed about collecting these figures and so suddenly the numbers change. Perhaps accidents involving bicycles have to be reported separately, and what was once a database "field" containing all accidents now contains only non-bicycle accidents. If we are not aware of this, we could be fooled. Visualizations can work with imperfect data like this, and they can help us spot them when we didn't know about them before, but each quirk needs to be explained.

9.3 UNCERTAINTY OVER TIME

Chapter 8 looked at uncertainty, and with time series data, there is the special problem of predicting what will happen in the future. Our best guess might be that trends will continue, but we do not know this for certain, so as time goes by, the number of other disturbances that might come into play increases and uncertainty should increase too. For simple line charts, this allows a nice way

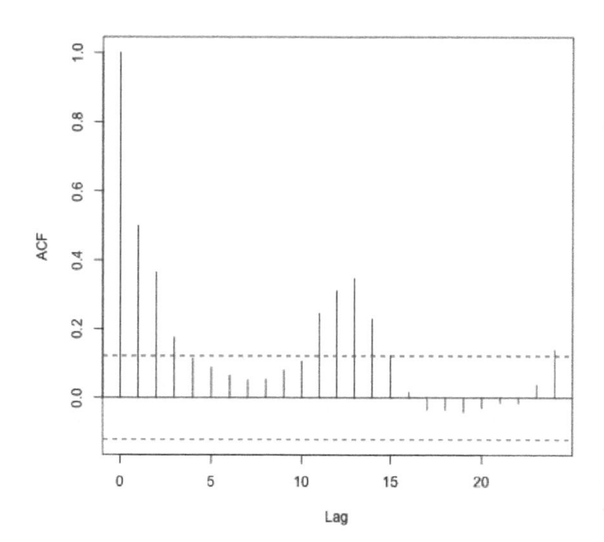

Figure 9.6 Autocorrelation plot for the train delays data shows a bump around a lag of 13 four-week periods which reflects annual seasonality. The blue dashed horizontal lines indicate the range within which we would expect to find fluctuations due simply to random noise.

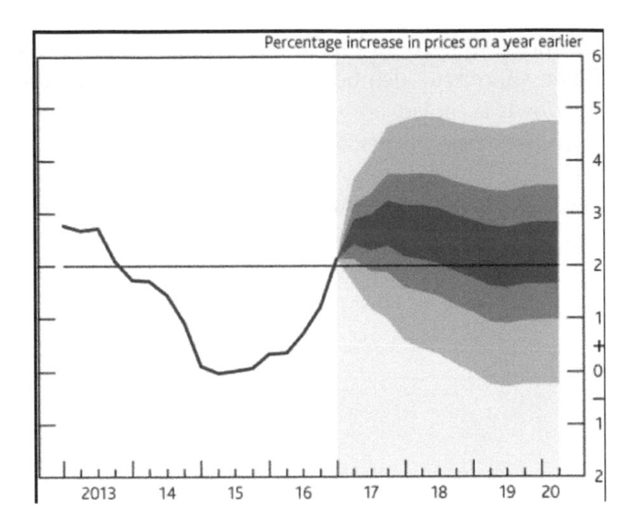

Figure 9.7 A fan chart showing observed (and therefore precisely known) past inflation, and future projections with increasing uncertainty. Copyright Bank of England, reproduced with permission.

of showing the predictive uncertainty by having a shaded area that fans out as we go into the future. These are called fan charts (Figure 9.7).

In the same way we can calculate confidence intervals, we can also get software to give us predictive intervals. This is like predicting where the hurricane will move to on the map. Shading is one way to visualize it, multiple lines is another, and we can also have a sharp boundary or fade out the fan at the edges.

9.4 STATISTICAL TIME SERIES MODELS

Statistical models can be fitted to our data to explain how the time series is comprised of a long-term trend, one or more periodic effects, and short-term noise. To continue the traffic accidents example, the long-term trend might be that roads are getting safer thanks to improved technology in cars and awareness among drivers. Each year there may be more traffic accidents in the winter because of wet and icy roads, and each week there may be more accidents on Mondays to Fridays because of the volume of traffic commuting to work. There may also be daily patterns reflecting rush hours, taking kids to and from school, and drunk drivers late at night. Then there will also be some unexplained ups and downs that we put down to noise.

Whatever patterns we discover, we need to consider whether we believe the explanation, and we need to show it honestly. Government statistics are often seasonally adjusted, which means that the analysts have tried to separate some periodic effect over the year from a longer-term trend of interest. Simply writing that it is adjusted, without further information, can arouse suspicion that the figures have been manipulated, but showing what it entails visually can reassure the reader that it is not only sensible but actually essential to understand the data.

In Figure 9.8 we see an example of decomposing a time series into different parts, which shows the daily number of births in the United States each day from 1969 to 1988. There is a lot of information here. The first chart shows a mean over all the data as a horizontal line (and sets that as a baseline, so everything else has a relative index where the mean is 100), a long-term trend and shorter-term noise. The second chart shows a consistent effect of days of the week: more babies tend to be born on week days. There is a line for each of 1972, 1980, and 1988, so we can see that this has become more pronounced over time. The third chart shows a smoothed seasonal effect within each year, and again the individual lines show that this has become more pronounced. Finally, there is an effect for each day of the year, which is not smoothed, and we can see the reduced births on significant holidays, with attendant increases before and after them. A little annotation of the holidays helps to tell the story.

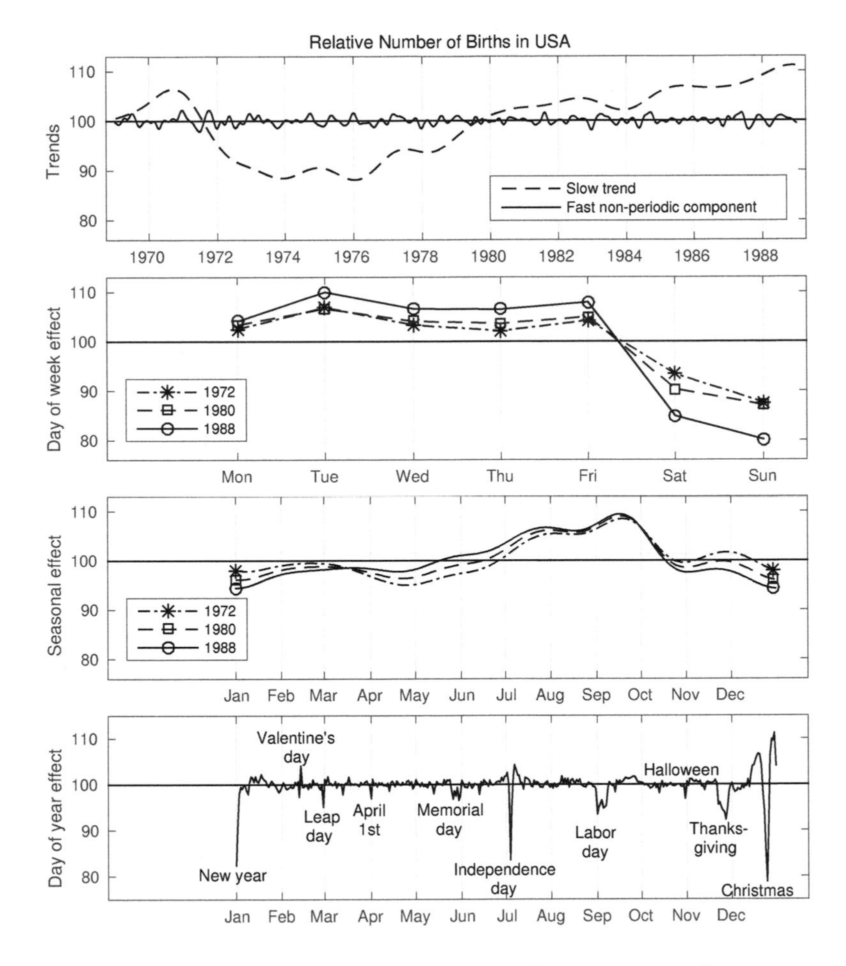

Figure 9.8 Time series data decomposed into seasonal components, slow and fast trends: daily number of births in the United States each day from 1969 to 1988. From *Bayesian Data Analysis* by Andrew Gelman and colleagues. Copyright CRC Press.

CHAPTER 10

Statistical predictive models

MORE PEOPLE IN MORE SETTINGS than ever before are interested in using data to predict the future. How many new sales will this marketing strategy produce? How will people vote in this neighborhood? What will unemployment be like over the next two years?

These are all applications of statistical models. This chapter will explore visualization aspects of regression models, while Chapter 11 covers trees and machine learning methods.

They all have some variables that are known and used to predict some other variable of interest. I will call these predictors and outcomes respectively, but you might also hear talk of inputs and outputs, independent and dependent variables, exogenous and endogenous variables, covariates and target variables. They all mean the same thing in simple predictive models.

In the machine learning community, this is called supervised learning, because the outcome is known and you can check the predictions against it to improve or supervise the model.

Having come up with a model that seems to predict the outcome quite well, it is tempting to omit all the details of how it was derived, and the assumptions it might rely on. A recurring theme in this book is the idea of layering different levels of detail so that readers can drill down to the extent they want. Don't hide the technical details completely – put them in a deeper layer of visualization and explanation.

10.1 LINEAR REGRESSION MODELS

The easiest predictive models to grasp are those with one continuous outcome variable. We have a dataset where the predictor and the outcome are both known for all observations, and we ask the computer to find a formula that best relates one to the other.

Later, someone can apply this formula to new data, when they know the predictor but not the outcome, by simply plugging the predictor value into the formula. We can extend this idea: there can be multiple predictors, and the formula can have a simple or complex form, and those choices are up to the analyst.

With just one predictor and one outcome, visualization is straightforward, because there are two variables and we can encode these as horizontal and vertical positions in simple, familiar formats. The data can be shown as a scatter plot, and then the formula is some kind of line or curve through the markers.

If we encode the predictor to the horizontal position, and the outcome to the vertical position, as is conventional, then the vertical distance from the curve to a marker is called the residual: a measure of how wrong the prediction is for that observation.

We can visualize both the observed and predicted values in a scatter plot, as long as we link them visually for each observation. This has an extra interpretation, that the vertical linking line shows the size of the residual (Figure 10.1).

The goal is to pass close to the markers (Figure 10.1), but not to create a tortuously complicated curve that does a great job with the current data and will actually do quite badly with the next lot – which is called over-fitting: (Figure 10.2). In short, the prediction should follow the overall trends and patterns and ignore the noise.

Cross-validation is a useful tool for finding optimal predictive models, and it also works well in visualization. The concept is simple: split the data at random into a "training" and a "test" set, fit the model to the training data, then see how well it predicts the test data. As the model gets more complex, it will always fit the training data better and better. It will also start off getting better results on the test data, but there comes a point where the test data predictions start going wrong (Figure 10.3).

The goal of the predictive modeler is then to fine-tune the model until it works well for both training and test data. It can be helpful to visualize some measure of model accuracy on the test data. In linear regression, the root mean square (RMS) error is a common

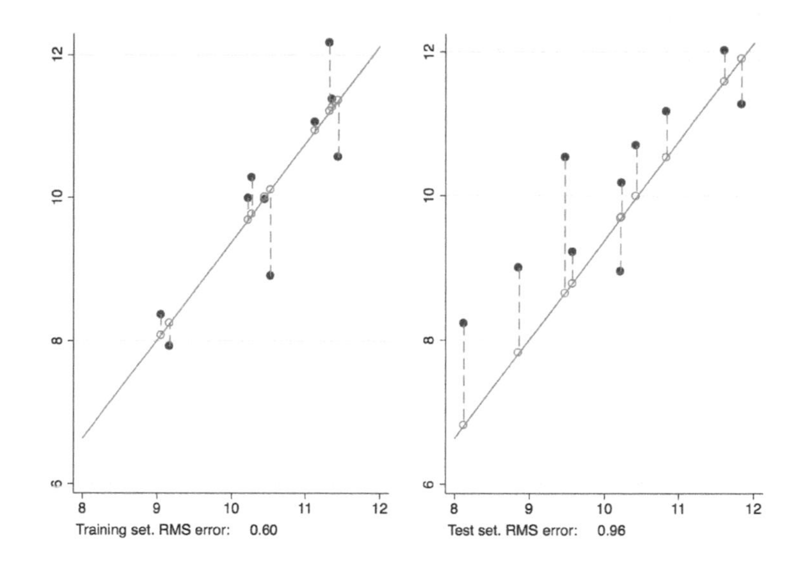

Figure 10.1 A simple linear regression model, with a straight line, fits both training (left) and test data (right) quite well, as shown by root mean square (RMS) error

choice. Different versions of the model will produce different RMS errors. We can compare them in a line chart or dot plot.

We might also want to divide up the data, to see if there are observations that are often poorly predicted, as this could prompt us to design the model differently. So, we might want to have an RMS error for each observation over multiple models and cross-validation slices of the data. We can then draw these against other variables and aspects of model complexity.

In Figure 10.3, we try predicting the vertical position (the outcome) with just a single mean value for all data, which actually does quite well and is of complexity 0. Then, we try including the horizontal position's predictor variable: a straight line, complexity 1. Then, the predictor squared, which allows a U-shape that the data possibly exhibits (complexity 2). Any of these could be accepted, but adding more complexity leads to a poor fit to the test dataset.

In the case of the straight line, we can talk about the slope of the line, which defines how much the outcome changes if there is a

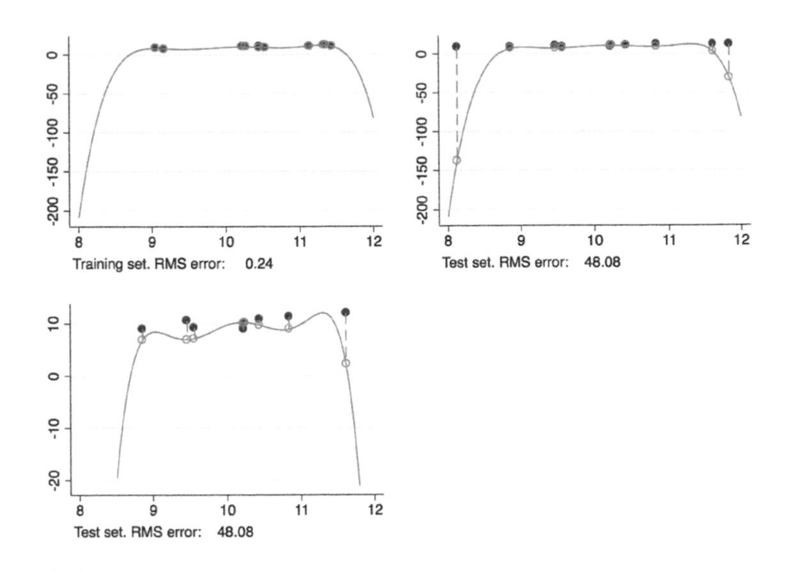

Figure 10.2 A more complex regression gets a closer fit to the training data but is over-fitting, because the test data are much farther from the predicted curve, as shown by the large RMS error. Even when we zoom in on the central region, (bottom left) the size of the errors is much bigger than the simple straight line achieved.

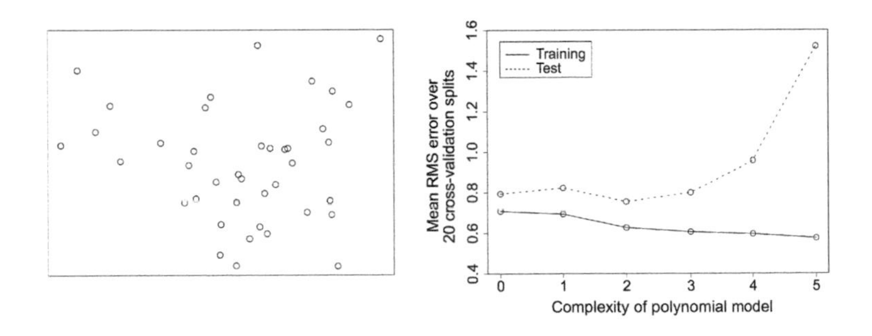

Figure 10.3 As the data on the left are used to fit increasingly complex regression models, the RMS error slowly decreases for the training set, but increases rapidly for the test set after a certain point.

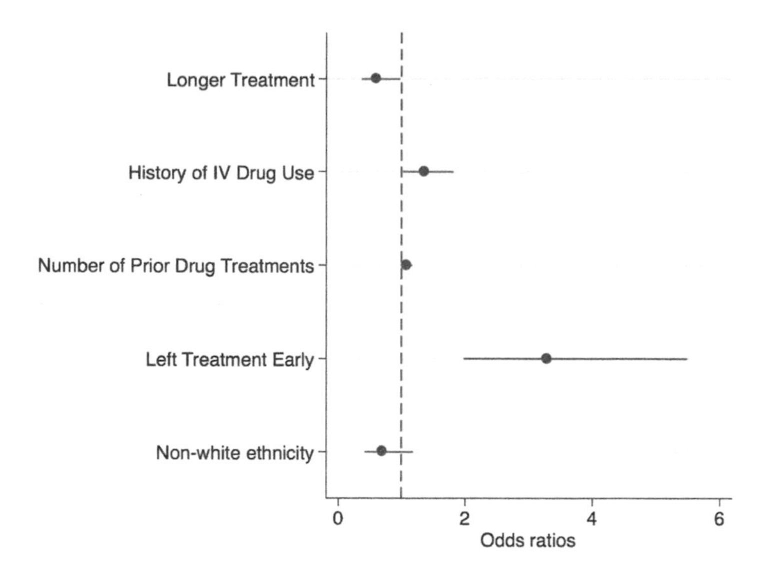

Figure 10.4 The odds ratios, and their confidence intervals, from a logistic regression for relapse in the IMPACT study, shown in a dot plot. The red dashed line shows an odds ratio of 1, indicating no effect. Confidence intervals lying entirely to the right of the red line indicate statistically significant harm, and those entirely to the left indicate statistically significant benefit. The borderline significance of the benefit of longer treatment is visible at the top, but beware of interpreting the other odds ratios as cause and effect.

change in the predictor. A steeper slope means a stronger "effect." We can simply show the slope, which is quantified by a regression coefficient, as a marker with error bars for its confidence interval (Figure 10.4). A positive slope indicates that observations with higher values of the predictor usually also have higher values of the outcome too. A negative slope means the opposite.

A slope relating a predictor variable to an outcome might not indicate a causal relationship. I can use sales of ice cream to predict the number of people infected by common colds, but the ice cream is not preventing colds, they are just correlated (see Section 6.2).

What if the regression has more than one predictor? If there are two predictors, you might be tempted to visualize them in a three-dimensional format with the outcome encoded to height, although this has some limitations (see Figure 12.3). Instead of a line, there will be a surface slicing through the three-dimensional space that gives a prediction of height for each combination of the two predictors. Of course, this is no help once you have more than two predictors.

We can show each predictor in turn against the outcome in scatter plots with curves for the predictions. Now, with a more complex model, the combined effect of all the predictors does not necessarily indicate a straight line in such a visualization, even if each one is itself just a simple slope.

Having a slope for predictor A and another of predictor B suggests that their effect on the outcome add together. Our models don't have to stop there, even in this simple linear regression method. We can have interactions between predictors, which allow the effect of A and B together to be more than (or less than) the sum of their parts: the A slope plus the B slope plus some other combined effect.

Interactions make visualization essential because we can no longer simply use the coefficients to summarize the model. If we visualized predictor A and the outcome together, we would have to state what value predictor B has been set to, because it changes the slope of predictor A. Often, people set them to the mean value in the dataset, but more sophisticated software can generate marginal effects and show the effect of multiple predictors in combination; we'll return to this below in logistic regression (Figures 10.5 and 10.6).

We have to be very careful about prediction out of sample, which occurs when we assume the relationship we have found in our model applies even beyond the range of predictor values we had when we got the software to fit the model. You can see this happening in Figure 10.2 where the prediction line does a great job inside the range of the training data but heads off to improbable values once it is outside that range. Visualization makes it obvious when we are straying outside the data.

10.2 LOGISTIC REGRESSION MODELS

If the outcome is binary, we can give it values of 0 (no) and 1 (yes). We wouldn't expect a straight line to go through these points in a

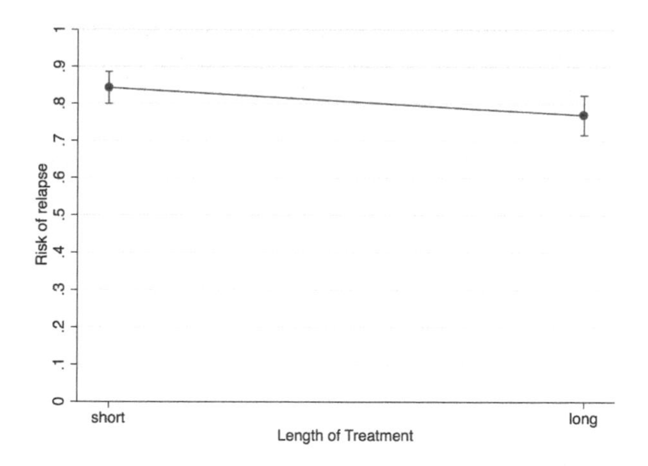

Figure 10.5 Marginal risks of relapse among IMPACT study participants in the two treatment durations, with 95% confidence intervals. Marginal risks are useful because all the complexity of the adjustments in the model are captured.

scatter plot, at least not well, so there is a different sort of curve that we commonly use, called the logistic function. It predicts the risk of the outcome happening, a value between 0 (definitely not going to happen) and 1 (definitely going to happen).

Unfortunately, to obtain this we have to have a slightly different formula using the predictors, and the coefficients are no longer slopes but rather odds ratios. We already found in Chapter 4 that odds ratios are not very easy for people to understand and discussed ways of converting them to something more easy to visualize and communicate.

Because the outcome values are either one (happened) or zero (didn't happen), encoding them as a position is not helpful. Too many of the observations would be piled up on top of one another to be seen clearly.

In Section 6.2 on correlations, I mentioned a study of residential courses of treatment for drug addiction. This was a real study called IMPACT, run by the University of Massachusetts Aids Research Unit, which looked at people attending two facilities in New England, and allocated them randomly to either a short or long course of treatment. However, the two facilities were very differ-

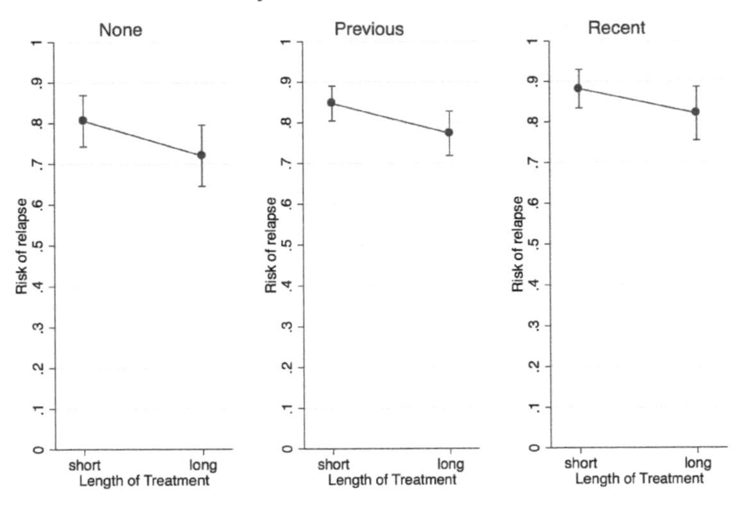

Figure 10.6 Marginal risks of relapse among IMPACT study participants in the two treatment durations and over three different types of previous treatment history. The fact that switching from short to long has almost the same effect in each group reflects the lack of an interaction in this model ("almost" because of the conversion of odds to risk).

ent, and actually the long course at one was the same length as the short course at another. They also took in different groups of people, who differed in their addictions and how many times they had tried to recover.

In an experimental study like IMPACT, the aim is to estimate the cause-and-effect relationship between treatment duration and risk of relapse. If it had been more rigorously conducted, and there had been no notable difference between facilities, we could simply compare the risk (or rate) of relapse between short and long treatments. A bar chart might show the two risks or rates, and a risk (rate) ratio could be calculated and shown as annotation. Or the ratio itself could be shown in a dot plot with its confidence interval. A waffle plot or pictogram might make this more concrete in terms of people staying clean or relapsing.

However, this study was complicated by those differences between the two facilities. In situations like this, the analysis must try statistically to separate the difference of interest (treatment duration) from the one mixed up with it (previous addictions and treatments, introduced via the facilities). This is called adjusting and most times you read that some figures have been adjusted, it will have been done by including those issues as predictors in a regression model. (We saw this with the Million Women Study in Figure 6.4 and in Figure 10.4.)

Suppose we include the treatment duration and the facility as two predictors. That will give us two odds ratios, one showing how much of an effect double-length treatment has, and the other showing the odds ratio between the facilities.

Only the first of these – the odds ratio for duration – is of interest and we say it has been adjusted for facility. We might want to visualize this duration odds ratio and its confidence interval, but there is not much point in visualizing the facility odds ratio. We can present it in a table or text instead.

Other variables can be added too, such as the number of failed previous treatments, or the participant's age. We call these confounders when they get in the way of a causal estimate: the effect that a change on duration has on the outcome. They are only in the model so that we can separate them from the effect of treatment duration, and should not be visualized in their own right (because they are *not* causal estimates of the effect of failed previous treatment, age, etc.).

If a confounder has an interaction with the predictor of principal interest (treatment duration in this case), then that means that changing treatment duration won't affect everybody's odds of relapse in the same way, because the effect of the confounder and the long treatment is not simply the two odds ratios multiplied together. For example, someone who tried and failed several treatments before might experience little benefit compared to someone who is trying to give up for the first time.

If we believe that to be the case, we can try including that interaction in the model. In such a case we would have to show the predicted risk changing over values of the confounder, like in Figure 10.6. This visualizes the marginal effects, which is a way of estimating what the effect of the treatment is, taking the other variables into account.

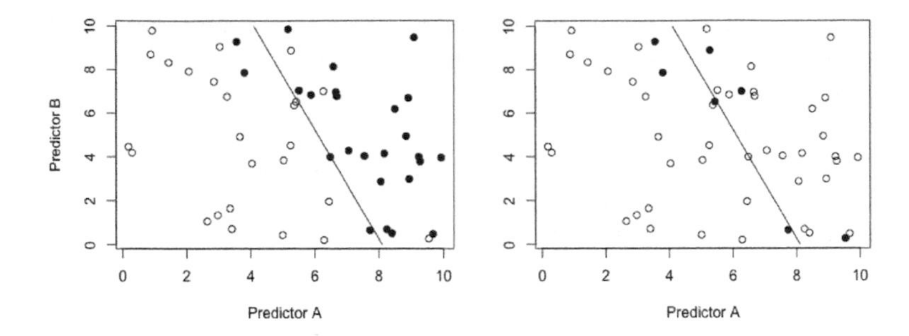

Figure 10.7 Left image: two continuous predictor variables and a binary outcome with a line showing where a logistic regression predicts a 50% chance of the outcome having happened (black markers). Right image: we can also show just whether the regression prediction is wrong.

Alternatives that you may encounter, such as partial dependence plots and accumulated local effects plots, look similar but use a different calculation underneath.

In this example, the treatment variable is categorical, but if it was continuous, like the dose of a drug, then we could have a line chart with shaded confidence region instead, or something similar.

Logistic regressions give us a predicted risk, from 0 to 1, and we then have to decide whether each case is predicted to have or not have the outcome, by imposing a threshold, over which we predict a "yes," and under which we predict a "no." Once we have done that, we can draw visualizations of both the old data used to fit the model, and any new data, with the predicted status encoded as some suitable parameter(s) like color, size or shape (Figure 10.7).

The line was placed in Figure 10.7 by finding the value of predictor A that gave a 50% risk prediction if predictor B was zero, and the same thing when predictor B was ten. This is essentially the location of the line at the top and bottom of the scatter plot. Then, a line can just be drawn between these two points.

10.3 SEMI- AND NON-PARAMETRIC MODELS

If you read about smoothing in Section 7.1, you might wonder how this is different from regression. In visualizations, it certainly looks similar: a curve moves through the data, trying to follow any patterns. Up to this point in this chapter, we've been looking at parametric models. They relate the predictor(s) to the outcome by a formula with some parameters, such as the regression coefficients we've already seen.

For example, in the IMPACT study, to adjust for the number of previous failed treatments as a confounder, we might take that number for each of the study participants, multiply it by a coefficient (which happens in the case of logistic regression to be the logarithm of the odds ratio) and then add that to the rest of the formula. Once we know the parameters, we have fitted the model.

This is neat and simple but requires us to choose what sort of shape the model curve will take up front, before fitting the model. An alternative is non-parametric regression, where we allow the data to determine the shape, perhaps using a smoother like splines (see Section 5.4). Predictions can be generated from the smoothed line or surface.

With non-parametric regression, there is no formula, just a shape that curves through the data, so here visualization really is essential. I discuss an interactive web page about splines in Section 14.3. Generalized additive models are one of the most popular non-parametric techniques nowadays; they allow there to be a complex, non-parametric curve for each of several predictors, and these added together make up a prediction.

The limitation of non-parametric models in data visualization is that we cannot draw coefficient plots like Figure 10.4 because there are no coefficients.

We can also mix these two together to get regression. Perhaps the effect of the number of previous failed addiction treatments is determined by a smoother, but that is then added to a parametric formula with odds ratios for the treatment duration and facility. Then the predicted risk of relapse would take each person's position on the smoothed curve and multiply it by the duration and facility odds ratios. If long treatments are helpful and therefore reduce the risk, the whole curve would drop downwards for people getting the longer treatment.

Time-to-event or survival data require special models, and one of the most common ways to examine them is with a Kaplan-Meier plot. These show the percentage of study data still "surviving" at various times. These needn't be literally dying people: the data could also be something like the time to first claim on an insurance policy, and the event in question could even be desirable.

As time went by, people in IMPACT either disappeared and lost

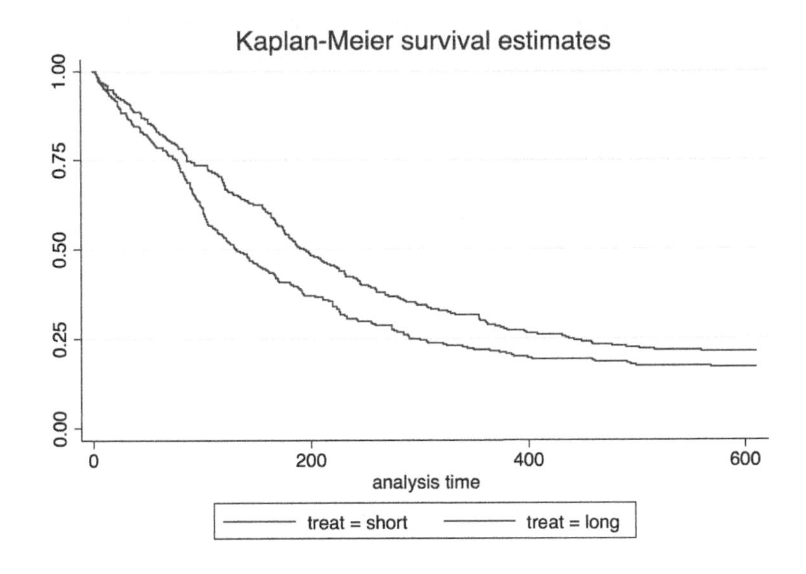

Figure 10.8 Kaplan-Meier plot showing the proportion of IMPACT study participants still at risk of relapse (therefore having "survived") at different points in time for the two treatment durations.

contact with the researchers, or were recorded as having relapsed and started taking drugs again. At the end of the study, anyone remaining in contact and drug-free is recorded as such. In most Kaplan-Meier plots, the survival curve is shown as a step, which recognizes the individuals leaving the pool of people at risk. Confidence intervals can be added, although there is scope for confusion when they overlap.

The survival curve seen in the Kaplan-Meier plot could be a strange shape. I once analyzed data on people dropping out from a long-running study, and treated it as a survival problem. Every couple of years, the study would send out questionnaires and invite people to medical examinations. It tended to be at those times that people asked to withdraw, so the survival curve looked like a staircase with a series of sharp drops.

For situations like this, it is common to use a semi-parametric model called Cox regression to see how other predictors, like the treatment duration in IMPACT, raise or lower that curve. Cox regression parameters are usually provided by software in the form of hazard ratios, which behave like risk ratios and can be drawn in a dot plot.

10.4 HOW GOOD IS THAT MODEL?

Any fool can fit a statistical model, given the data and some software. The real challenge is to decide whether it actually fits the data adequately. It might be the best that can be obtained, but still not good enough to use. With linear regressions, we can calculate residuals for each of the data points by subtracting the observed outcome from the predicted outcome.

A positive residual indicates an under-estimate, and a negative one an over-estimate. The residuals can then be drawn in a scatter plot against the predictions of the observed values, and possible predictors that have not been included in the model. A good model will have a uniform cloud of markers with no discernible trends to them in each case.

With logistic regressions, the residuals are not calculated by simple subtraction. Instead, we are interested in whether the predicted outcome (obtained by imposing a threshold value on the predicted risks) and the observed outcome differ.

The location of the threshold matters here. Cutting at a risk of 0.5 (50%) might seem the obvious choice, but it is not guaranteed to give the best predictions. That would mean that any predicted risk under 50% means predicting that the outcome will not happen, and any risk of 50% and above means it will happen. As we increase the threshold, we will predict more data as "no" and fewer as "yes." This will mean more are false negatives and fewer false positives. Lowering the threshold has the opposite effect.

A common, if not very intuitive way of showing this balancing act is by drawing a receiver-operator curve (ROC). This is a line chart with the proportions of the two types of wrong prediction encoded to the two axes, and tracing out what happens as the threshold moves from 0 to 1. A good model will have a line that bulges upwards, while a model no better than random guessing will have a diagonal line. The area under the curve (AUC), as a proportion of the whole chart, is one way of assessing the overall fit of the model, indicated by a high AUC.

10.5 USES OF COMPUTER SIMULATION

Bootstrapping can be used to obtain estimates of uncertainty in regression models, and the resulting many semi-transparent curves through the data may be easier to read than shaded areas on either side of the best guess.

Cross-validation also works by getting the computer to simulate many analyses. If you have enough data, you can split them into a training set and a test set. Fit the model using the training set and then assess its fit on the test set. As the model becomes more complex, it will always get better at fitting the training set but after an optimal point, will actually start to get worse on the test set. This is the over-fitting mentioned earlier. If this split-and-fit is done several times, the results can be averaged and visualized for increasing model complexity.

You might, for example, consider adding interaction terms to a regression and assess that decision with cross-validation. We need to communicate the choice of model as well as its results, and this is a good way of bringing that out into the open while also being quite intuitive. Line charts can show some measure of error for training and test datasets as the model complexity increases (Figure 10.3).

Another useful approach to testing models using simulation is to generate phony data using the estimated parameters of the model. For example, if you get a line through the data, with a certain amount of scatter above and below the line, simulate new data along that line with that scatter. Then compare the phony data to the real data. If there are discrepancies, it may highlight somewhere that your model can be improved. Visually, having the same charts of the real data alongside the phony data makes it easy to spot problems. This is sometimes called posterior predictive checking, especially in Bayesian statistics.

CHAPTER 11

Machine learning techniques

DATA SCIENCE IS OFTEN DESCRIBED as a meeting of methods to understand data, some from statisticians and some from computer scientists. The computer scientists generally refer to their techniques as machine learning. Newcomers to this can find the distinction between machine learning and statistics confusing, because there is not much more of a difference than just the choice of words. Some methods, like principal component analysis and logistic regression, get claimed by both sides.

These methods often help prediction with data where the patterns are not simple. This can make visualization difficult, but it also means that sometimes visualization is very valuable for understanding the model. Once you become familiar with the goal of the analysis, and the many examples of different visualizations that have been effectively used before, you will start to see connections: how an idea from one application could be brought to bear on a completely different problem.

The concept of marginal effects, which we encountered in Chapter 6, is widely used for these models too, although a line chart of the marginal prediction versus one predictor variable is often called a partial dependence plot in machine learning.

In this chapter, I won't attempt to give more than a very broad introduction in intuitive terms, and to highlight some important emerging areas that affect visualization. If you haven't read Chapter 10, and are unfamiliar with regression models, I suggest you read that first. I introduced cross-validation there, which is very useful to tune these techniques so that they perform well.

Also, I must point out that you may see some of these techniques described as "artificial intelligence," which is overblown to say the least. They may yet prove to be important tools for AI, but in themselves they are just building blocks. Membranes are vital for the functioning of our brains, but it would be foolish to refer to a sausage as intelligent for that reason.

11.1 ENSEMBLES, BAGGING AND BOOSTING

A recurring theme in machine learning is combining predictions across multiple models. There are techniques called bagging and boosting which seek to tweak the data and fit many estimates to it. Averaging across these can give a better prediction than any one model on its own. But here a serious problem arises: it is then very hard to explain what the model is (often referred to as a "black box"). It is now a mixture of many, perhaps a thousand or more, models.

For visualization then, we would have to abandon anything based on model coefficients, and work simply with predictions and residuals. When predictors include the position in space or time, this can be capitalized on with a map or appropriate encoding of the time variables (Figure 11.1).

It is tempting to visualize the prediction from each of the models that make up the ensemble, but this could be misleading. They are not intended to stand on their own as predictions. Some will work well only for a small subset of the data. Some will contain predictor variables (which machine learning people often call "features") that are absent from others.

Nevertheless, we have predicted and observed values of the outcome, and from those we can get residuals. It may be informative to look at the residuals and find what sort of observation is not predicted well. For those observations, we can then unpack the constituent models and see if there are some that perform better than others. This could help us open up the black box and understand how to improve our model. There could be many observations and many constituent models, so the only way to do this is with visualization. Simple approaches like line charts will work well, as the eye is drawn to anomalous spikes.

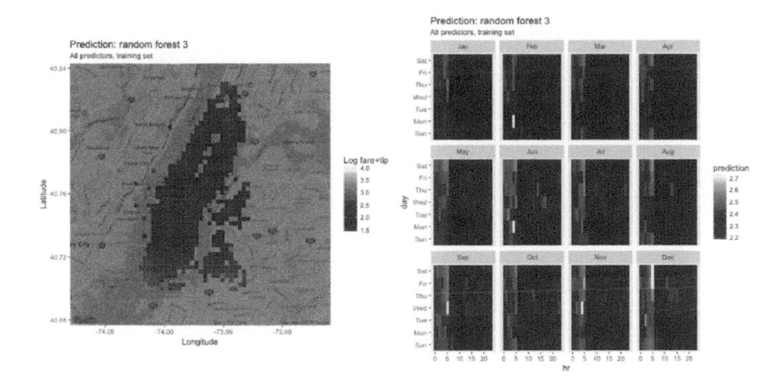

Figure 11.1 Predicted taxi fares and tips from a random forest model (explained below), fitted to a sample of journeys in New York. Latitude and longitude are captured in a map, while hour of the day, day of the week, and month of the year are in a heatmap that reflects these variables' cycles within cycles. A slightly higher prediction at the southernmost tip of Manhattan (the financial district) is the only obvious pattern in space, while a strong pattern in the time variables correctly reflects the way fares change between night and day, and a preponderance of early-morning journeys to the airports.

11.2 CLASSIFICATION AND REGRESSION TREES

In Chapter 10 we looked at predictive regression models, which always involved some kind of curve trying to pass through the observed data points. But we could also chop up the predictor variables at specific values, creating a sharp step from one prediction to another.

Trees, which can be used for categorical outcomes (classification trees) or continuous outcomes (regression trees), look at each predictor and judge how good a prediction would be if it simply split the data at some threshold value. The predictor that does best is kept, and the data split, like branches of a tree. Then it repeats that inside each of the branches. Eventually, not much can be gained, and some clever limits or "pruning" of branches ensures that it is not over-fitting (see Chapter 10 for more discussion on over-fitting).

This means that each observation belongs to some branch, and every observation in the branch gets the same predicted value.

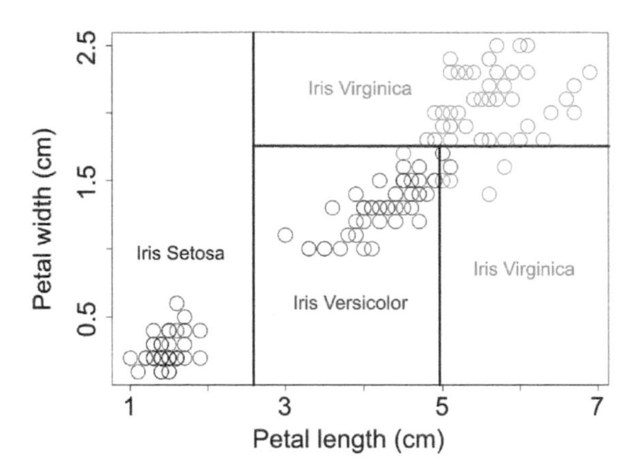

Figure 11.2 A tree classifying three species of iris flowers, on the basis of petal dimensions, shown as subdivisions of a scatter plot.

For binary outcomes, we predict that it either happens or doesn't happen in a particular branch; for other categorical outcomes, we predict which category every observation in the branch is most likely to belong to; and for continuous or discrete outcomes, we predict a value.

So, rather than a smooth curve, trees provide jumps to different predictions. There are two easy ways of visualizing this that are arguably easier to understand than regressions, though they only help in simple cases. First, if we have two predictors, we can draw a scatter plot with them encoded to horizontal and vertical, encode the outcome to color or size, or some other attribute of the markers, and then draw lines sub-dividing the scatter plot with each of the branch divisions. In Figure 11.2, the shapes of iris flowers are subdivided on the basis of petal length and width.

In each of the branches, which appear as blocks, we can show the predictive accuracy. One good option would be to annotate the block with a measure of how well the tree does.

For binary outcomes, a classification matrix showing predicted yes/no and actual yes/no would give complete information, although you might prefer a more compact measure like % correct. For continuous outcomes, you can summarize the residuals, either with something visual like a miniature histogram or statistically.

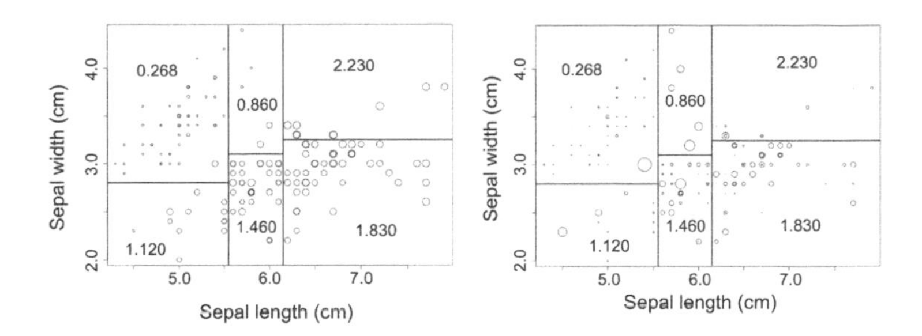

Figure 11.3 A regression tree, predicting petal width on the basis of sepal dimensions. Petal width (the outcome) is encoded to the marker size in the left image, and the residuals are encoded to the marker size in the right image. The large residuals in the corners of some blocks suggest that a tree might not be the best way to classify these data.

Just as we could have marginal densities on a scatter plot (Chapter 3), we could also add marginal plots of residuals versus predictor values (perhaps with a smooth regression line like LOESS through them). A random scatter (with the blue dot at the location of the branch) would indicate that the tree found reasonable points at which to branch, while a change at the point of the branch would suggest that maybe trees are not a good way to model these data.

Another option would be to highlight the (hopefully few) incorrectly classified observations. In Figure 11.3, the predicted petal width is shown in text in each block of the scatter plot – each branch of the tree. We can see on the left how it captures the overall pattern of the outcome. But on the right, we see the largest residuals around the diagonal. Trees, by splitting on one predictor variable at a time, fail to draw diagonal lines and can only approximate them by lots of little branches. This image shows that a linear regression would be better suited to this prediction.

Secondly, even with many predictors, we can show a decision tree, listing the thresholds for each split and the accuracy of the prediction against each branch. The scope for adding more small visuals is sadly reduced, though. There are no margins because no predictors are encoded to the horizontal or vertical positions, and with more than two predictors, there is no longer a single scatter plot that we can draw.

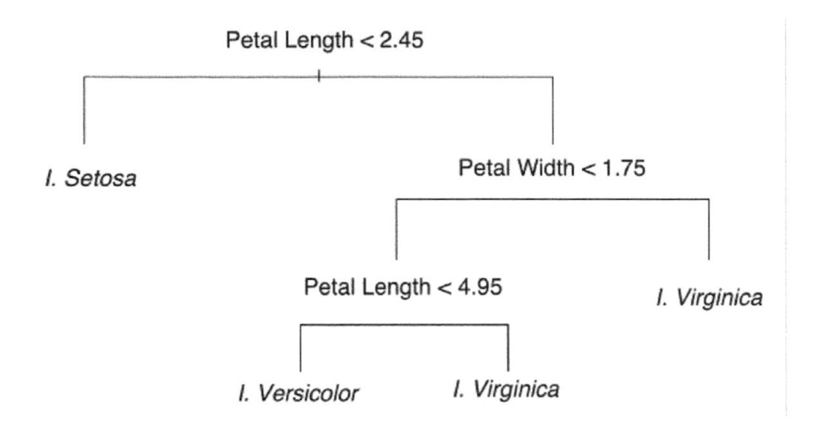

Figure 11.4 A decision tree diagram for Fisher's iris flowers.

Sometimes a tree has so many branches that it will be too wide to draw. But it may help to have the branches radiate out from a central point instead. This has been popularized by genetics, where having too many "leaves" on the tree is a common problem for visualization.

11.3 RANDOM FORESTS

Random forests are essentially an ensemble of trees. They use many short trees, fitted to multiple samples of the data, and the predictions are averaged for each observation. This helps to get around a problem that trees, and many other machine learning techniques, are not guaranteed to find optimal models, in the way that linear regression is. They do a very challenging job of fitting non-linear predictions over many variables, even sometimes when there are more variables than there are observations. To do that, they have to employ "greedy algorithms," which find a reasonably good model but not necessarily the very best model possible.

Trees, for instance, consider all the variables and partition the data based on the best choice. Then they look inside each of the branches. But the best overall prediction might not be achieved by picking the best branching first, because that has repercussions on all the branching that follows.

This is like how I play chess. I'm not very good at all. I try to look at all the possible moves I could make next, and take the one I think is best. But sometimes to win, you have to look farther

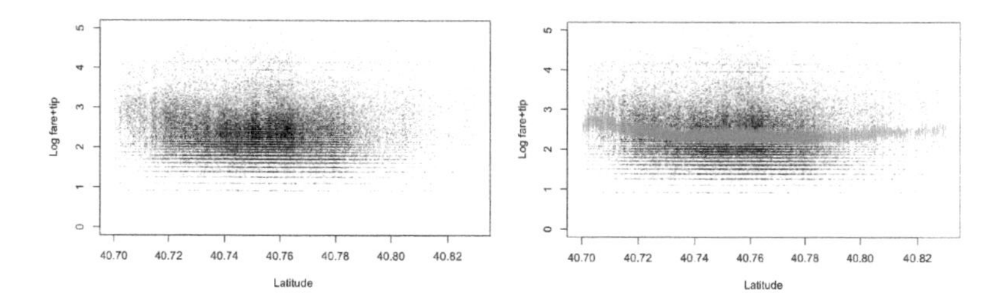

Figure 11.5 Latitude versus the logarithm of fare plus tip in a scatter plot (left); the same chart with a green line for latitude versus predicted log fare plus tip (right). The financial district effect (see Figure 11.1) is just slightly visible as an upturn on the left of the chart, and this is captured in the predictions (green line). Note that latitude is not the only predictor, so the green line jumps around as other variables like time of day push it up and down.

ahead, and make an apparently less good move now in order to set yourself up for the killer in a few moves' time. By averaging over all the trees, and altering the data, random forests jump past this trap.

Again, we have a black box. Each tree will lead to a different layout in a decision tree visualization, so we cannot simply do something like superimpose them with semi-transparency. We may be able to draw the variables as markers and the path taken from one branch to another, to see how often they follow in a certain sequence.

We could also get the computer to crank out, for each predictor, a scatter plot of observed outcome versus predictor, with a line chart of predicted outcome versus predictor value superimposed. If there is a pattern worth communicating to help people understand what the model is doing, it should show up in these (Figure 11.5).

It's also possible to obtain measures of how influential each predictor variable is on the final prediction, either as a standard output from the software, or in more heuristic fashion by altering the predictor values one at a time and seeing how much the prediction changes, and we can easily visualize these like in Figure 10.4 or use them to order the predictor-by-predictor visualizations. Each

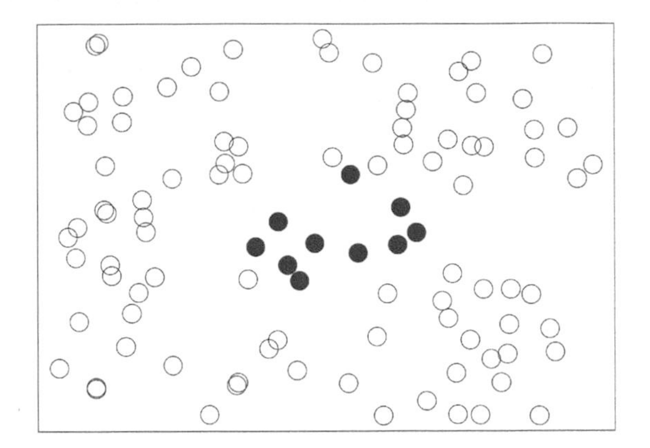

Figure 11.6 Regression models like logistic regression can struggle to correctly classify data like this, where there is no straight line separating the two types of point. Another example is the interlocking spirals in Figure 11.7.

case will be different when we have an ensemble of models, and visualizing them is an area where experimentation is still going on.

Because the "forest" is a collection of steps between variables (one is used to branch, then another, and so on), we may be able to visualize it as a collection of lines connecting variables, using some of the network techniques in Chapter 13. Whether this is successful depends on the complexity of the trees and variability between them; often the forest is comprised of very short trees.

11.4 SUPPORT VECTOR MACHINES

We looked at logistic regression in Chapter 10 as a way of separating our variables into a region where the outcome is likely to happen and a region where it is not. Logistic regression does that with a line or surface, but some datasets have more complex patterns that are not so easy to separate.

What if the outcome tends to happen in a small range of predictor values, and outside that, whether higher or lower, it does not happen? This might look like the scatter plot in Figure 11.6.

No straight line will successfully split the data and predict the outcome.

However, if the predictor variables are replaced with distorted versions of themselves, this will be possible. You could imagine taking the scatter plot, printing it on a sheet of fabric and then pulling the central region of the fabric up.

This is what support vector machines do with a lot of computer-intensive calculation, and then you can predict that the outcome would happen in any part of the fabric above a certain height. Rather than using horizontal and vertical positions, we are using a (potentially complex) combination of them.

With methods like support vector machines or neural networks, we cannot employ the straight line we added to a pair of predictor variables in Figure 10.7. There is no straight line dividing up the data. Instead, we could try to find pockets of particular predictions by getting predicted outcomes for a grid of points and then making a contour plot.

This general concept of stretching and distorting the space that the original predictor variables define is a powerful one, but it causes a headache for visualization. The dilemma is to choose between showing the predictor variables or the stretched and re-combined versions of them. The original predictors which might relate to the outcome in such a complex and non-linear way that we can no longer see what the relationship is. But on the other hand, the combinations of them that are used to identify pockets of a particular outcome do not have any real-world meaning, like the predictors do.

It also gets used to help visualization in a method called t-SNE, which we will encounter in Chapter 12.

11.5 NEURAL NETWORKS AND DEEP LEARNING

Imagine you have a batch of logistic regressions, each with the same predictors, but different coefficients so they give different predictions. Then build another batch of logistic regressions, taking the predictions out of the last batch as *their* predictors. Continue like this and eventually boil it down to one prediction.

This will be a very complex model, capable of non-linearities that pick out little patches in the data. Because the cascade of connections forward from one batch of regressions (now you know the structure, let's call them layers) to another was originally conceived of as an imitation of neurons in the brain, this is called an artificial neural network.

They used to be too hard to compute beyond the simplest of examples, but recent improvements in software, hardware and cloud computing have made them possible even with a "deep" structure of many layers; more than three layers is called deep learning.

They don't have to involve logistic regressions as such, and they don't have to have connections between every "neuron" in adjacent layers. One variant, called convolutional neural networks (CNNs), has become very useful with image data, and because images lend themselves directly to visualization, I'll give CNNs their own section below.

Are they a black box? They are certainly harder to grasp than a simple regression, but the fact that they imitate physical structures makes them easier to visualize than many other predictive models. In theory, we could work out a formula based on the cascade of coefficients in each layer, but it would be so complex it would give us humans no new insights. However, plotting predictors versus observed and predicted outcomes, as suggested for random forests above, can give insight into the most informative predictors, and what sort of pattern the model is suggesting for it.

If we can draw the network and also draw something about predictions, then we could have a small visualization at each neuron, which helps us see how the prediction is built up, and if it is going wrong somewhere, where that might be. Figure 11.7 is a screenshot from an interactive web page that does just this.

There are two predictors and the outcome is binary, shown by color. The data are shown with circular markers and the task of the algorithm is to color the background so as to match the pattern in the data. You can choose one of the datasets on the left, select from seven features (machine learning terminology for predictors or functions of predictors), and then press play to let the software try to find a solution. Layers and neurons within layers can be added and removed. By hovering over a neuron you can see what its contribution is to the prediction, and by hovering over a connection you can see its weight (influence on the next neuron it connects to), also visualized by the color and thickness of the curve.

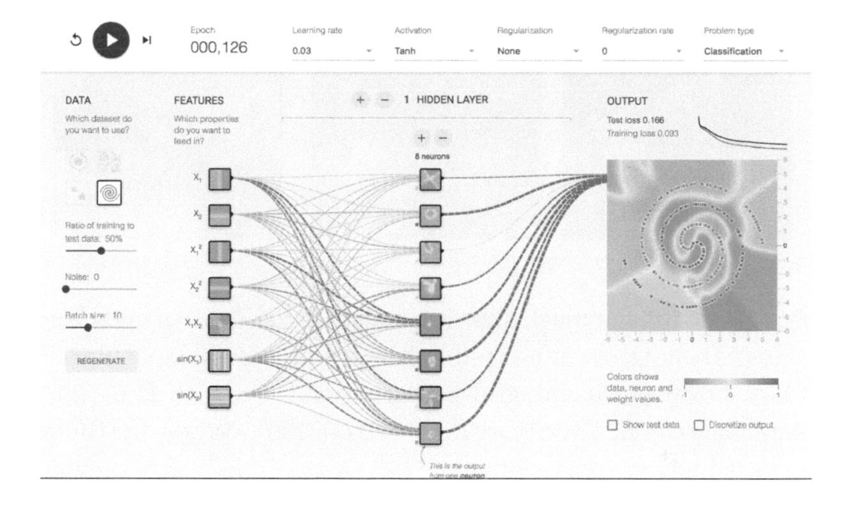

Figure 11.7 TensorFlow playground, an interactive illustration of how neural networks fit predictions to data. By Daniel Smilkov, Shan Carter, Martin Wattenberg and Fernanda Viégas at Google, used under the Apache 2.0 License.

11.6 IMAGE DATA

Convolutional neural networks (CNNs) are especially useful for analysing image data, where there is a two-dimensional grid of pixels. The pixels are the predictors, but only in relation to their neighbors. CNNs detect increasingly complex patterns in the images as the layers go on. The first layer might just predict yes or no on the basis of whether dark and light boundaries between dark and light patches appear, then in a later layer those will be combined into lines and curves with various orientations, then those get put together and start to detect recognizable shapes like faces.

CNNs are successfully used to automatically label images online for web searches, for example, whether the image contains a dog or a chair. Self-driving cars use CNNs to scan video cameras and identify hazards and information on road signs.

Figure 11.8 shows images from a recent study. The image on the left is classified by a CNN correctly as containing a flute with 99.73% probability. They then train the computer to blur parts of the image (center image) until it finds a place that destroys the CNN's chances of finding a flute. This is then the location of the

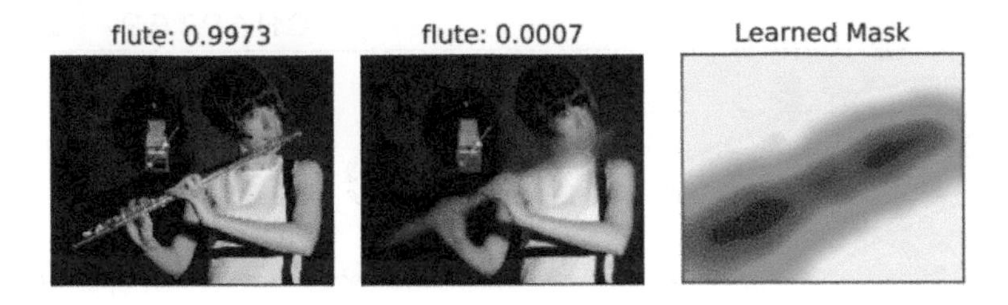

Figure 11.8 Image mask showing what a CNN detects and classifies as a flute, from the paper "Interpretable explanations of black boxes by meaningful perturbation," by Ruth Fong and Andrea Vedaldi (arxiv.org/abs/1704.03296). When the flute area is blurred (perturbing the data), the CNN's accuracy at detecting the presence of a flute drops from 99.7% to less than 0.1%.

flute, or at least the things that the CNN classifies as indicative of a flute (right).

Visualization is helping us to interpret the supposed black box. If we can see what the CNN is picking up on, we can check that it is being sensible. It is the closest equivalent we can get to examining residuals in a regression model. There is an infamous story of a military application of neural networks that had been trained with many photographs, some containing tanks and other not. The intention was to have automated tank detection cameras. The programmers got great results on these photographs and very poor results in field tests. Only later did they realize that every tank photograph had a cloudy sky, and many of the non-tank photographs had sunny skies – and it was this that the model was using to predict the presence of the tanks. With a visualization like Figure 11.8, they could have avoided the embarrassment.

CHAPTER 12

Many variables

HAVING TO SHOW COMPLEX INFORMATION in two dimensions is just a fact of life in data visualization. Statisticians call data with more than two variables multivariate. When there are more than two variables to be shown, we have to choose from:

- leaving some variables out,
- encoding some variables to less than perfectly perceived visual parameters like color,
- creating multiple visualizations, or
- trying to project as much of the information as possible onto a two-dimensional image.

This chapter will take a look at the third and fourth options. Of course, animation and interactivity can help, which appear in Chapters 9 and 14 respectively. There is always some compromise involved in showing it in two dimensions. Compromise is fine, as long as the strengths and weaknesses are known and any caveats are given clearly to the reader.

The individual variables should be described too, not just their combined complexity. Small multiple kernel densities will often suffice to give a quick overview of each variable that goes into a more complex visualization.

12.1 SMALL MULTIPLES

At several places so far in this book, I have alluded to small multiples as a way of showing more variables. The principle is simply

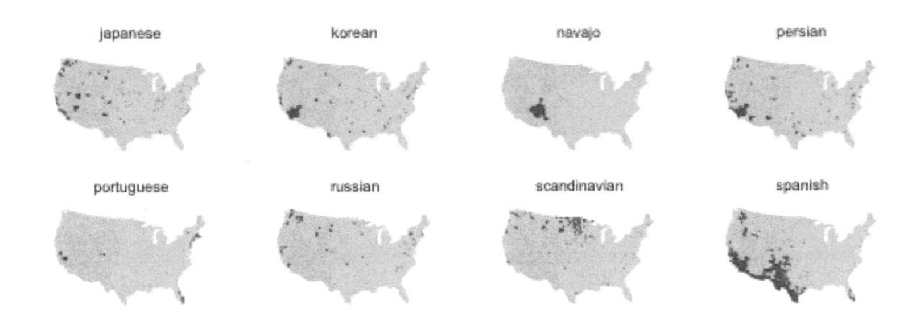

Figure 12.1 Prevalence of various languages in the United States, detail from a small multiple choropleth map by Reuben Fischer-Baum, used with permission. See Chapter 13 for more on maps.

to have multiple images that are identical except that different data are shown or highlighted in each. To avoid clutter, labeling and axes are usually not repeated but relegated to the side of the whole visualization. As a result, it may be that only very general patterns can be seen, but that might be the best available option. Small multiples can concisely show multiple variables, and maps are also well suited (Figure 12.1).

To allow comparison along the horizontal axis, small multiples will need to be positioned above and below one another, and likewise for the vertical axis they will need to be to the left and right. You can compare the ease of up-down and left-right comparisons in Figure 13.9.

12.2 IMPRESSIONS OF 3-D

If we have three continuous variables (or discrete or ordinal variables with enough distinct values), we could attempt to show them as if they were in 3-D. The human brain is quite good at perceiving things as 3-D even when they are not, so we can tap into that. Unfortunately, though, our visualization will not really be in three dimensions, so the reader will not be able to rotate it, walk around it or do any of the other things we do to understand the shapes of objects in the real world. So, there will be scope for confusion.

Consider the 3-D bar charts offered by many spreadsheet packages and adopted gleefully by beginners in data analysis (Figure 12.2). It may not be clear how tall the bars are when they do

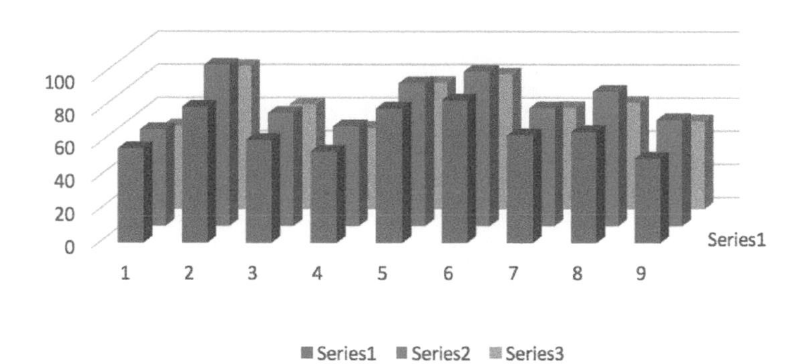

Figure 12.2 One of the standard presets for rows and columns of data in a popular spreadsheet package. This 3-D effect that is totally unrelated to the data is never a good idea; for example, can you tell if the blue in column 7 is taller than the gray?

not all start from the same line on the page or screen. Also, some will obscure others, affecting the first impression that a reader has (remember they will probably only look at a visualization for seconds).

In Section 3.4, we had some alternatives to scatter plots, that showed the density of observations at various points on a 2-dimensional surface (Figure 3.10). It is worth looking back at those as they can also be used to show the value of a third variable at locations on the surface.

We can draw a wireframe: imagine taking some chicken wire (a square grid) and shaping it into the surface you want, rising up from the first two dimensions into the third one (Figure 12.3). We can get software to emulate this and often readers will perceive it as a sufficiently realistic 3-D surface to get an immediate impression. It's important, though, that they do not see through the surface to what is behind it, or confusion will result – but then it is good to avoid one side obscuring another anyway.

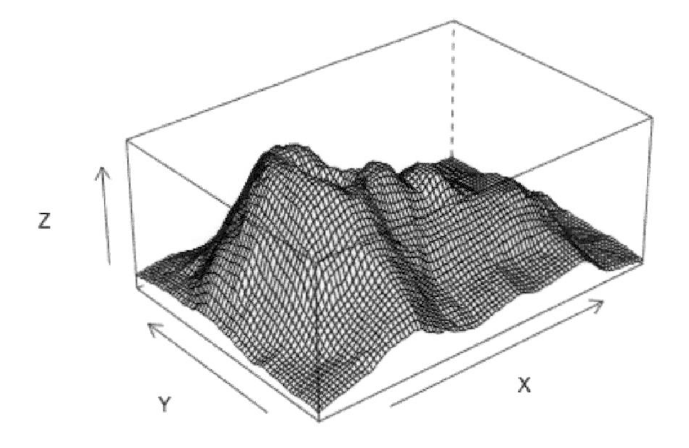

Figure 12.3 A wireframe plot of three-dimension data (three variables): latitude, longitude and height over a grid on the New Zealand volcano of Maunga Whau. Unfortunately, some of the data are obscured – a common problem with wireframes.

12.3 DISTANCES

A lot of statistical analysis is to do with distances. This is all to do with how "far" one observation is from another. Just as you can immediately see neighboring observations near one another with two dimensions in a scatter plot, we can define distances in more abstract ways. This leads to the idea of a distance matrix: a table of all the distances in the data, between any pair of observations.

Figure 12.4 is a distance matrix from a study of tuberculosis in the eye that I worked on, with data contributed from 26 hospitals around the world. There are 962 rows and 962 columns; each one belongs to one of the patients in the data. Where one patient's row meets another's column, the pixel is colored according to a statistical distance between them. The diagonal line appears because each patient is identical to themselves.

The rows and columns are ordered by hospital, and the fact that we can see a pattern like tartan cloth indicates that there are substantial differences in the data from the different hospitals. That might indicate different populations of patients, or different strains of disease, but it might also mean that they simply have different habits in recording the data. Without visualizing this, it would be hard to discern.

Statistical distances also play a role in putting things into the

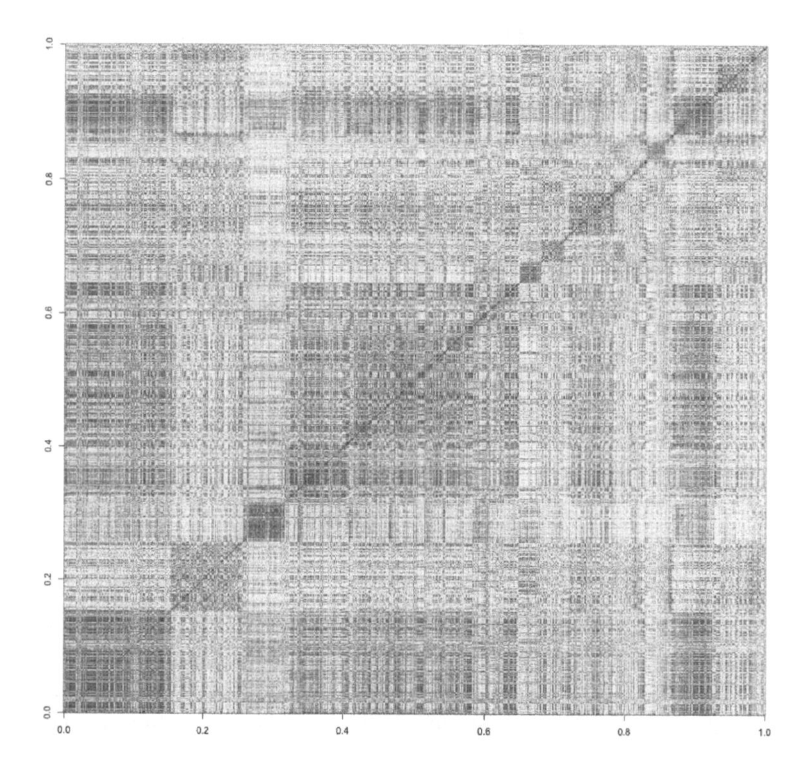

Figure 12.4 A heatmap of the distance matrix between 962 people with suspected tuberculosis in the eye. Red pixels indicate pairs of people with similar characteristics, yellow less so. The fact that blocks of color are visible suggests that different hospitals contributing data had differences in their patients, or in the way they recorded the data.

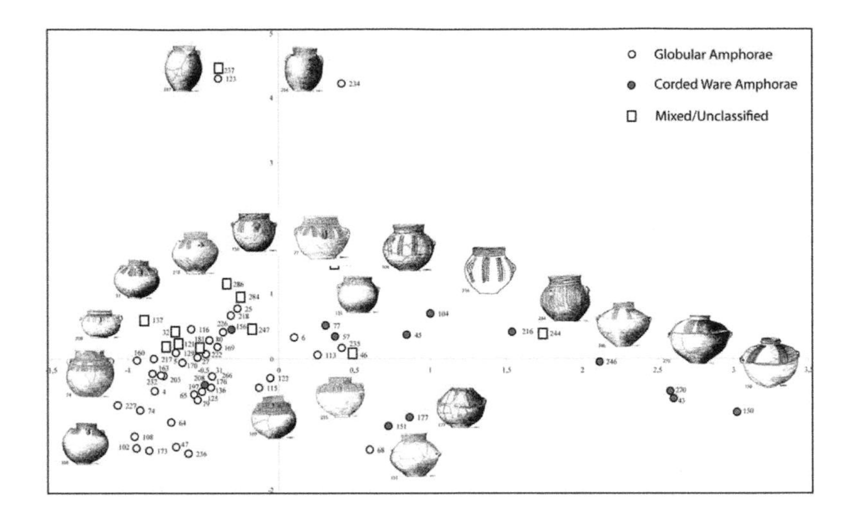

Figure 12.5 Multidimensional scaling allows archaeological finds to be placed in a two-dimensional scatter plot that best summarizes their differences, in this case the features of European Stone Age pottery. Apart from three unusual finds at the top, they mostly lie along a continuum from plain and round on the left to flatter and more decorated on the right. The archaeologists who drew this regarded it as evidence for a continuum of pottery forms rather than two distinct cultures. Annotating the plot with drawings of the objects is a helpful and engaging addition for readers. From *Upending a "Totality": Re-evaluating Corded Ware Variability in Late Neolithic Europe*, by Martin Furholt. Copyright Cambridge University Press, reproduced with permission.

right order. In archaeology, it's important to know what chronological order different artefacts were created in. If we dig up three ceramic pots and find A and B have six features in common, B and C have five in common, but A and C have only one, then they are most likely to have been made in order A, B, C or C, B, A (either in time or space). As soon as we have a distance like this, we can use a technique called multidimensional scaling to summarize it in two dimensions and then draw the points (Figure 12.5).

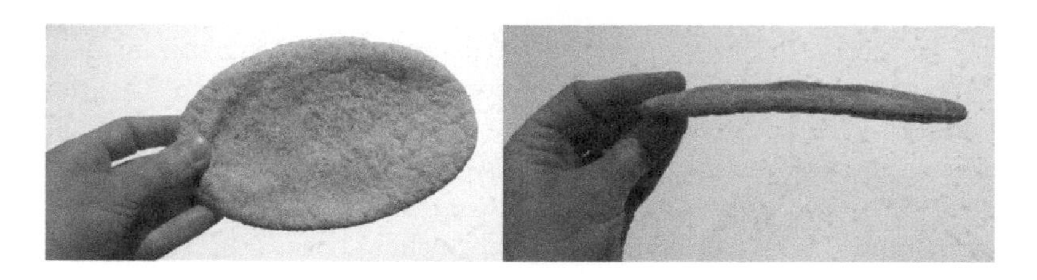

Figure 12.6 Data in three variables with different standard deviations can be visualized in a three-dimensional shape like a pitta bread. There is no ideal side from which to photograph it, to convey its shape, but the image on the left is preferable.

12.4 PROJECTIONS INTO TWO DIMENSIONS

Imagine a three-dimensional cloud of data shaped like a pitta bread. The three axes have different variances. If you wanted to take a photograph of the pitta and show people what it is like, it would be unhelpful to take the photograph end-on so that it looked like a long pencil-shaped finger of bread (Figure 12.6, right). You'd be losing a lot of information. However, even taking the photograph so that the two dimensions with the largest standard deviations are visible (Figure 12.6, left) fails to tell us how thick it is.

This is the idea of projecting multivariate data, which starts off with as many dimensions as there are variables onto fewer (hopefully two) dimensions (Figure 12.7). As soon as you go above three variables, you will not be able to picture this situation, but the principle is the same. We want to project the dots from the right angle to get insights into their structure.

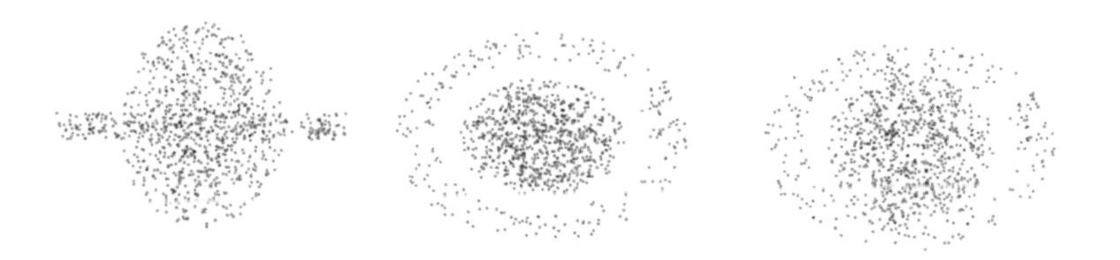

Figure 12.7 Three projections of a three-dimensional distribution of data into two dimensions. No single projection conveys all the information that this is shaped like the planet Saturn.

In Section 12.3, we encountered distances between observations. We can use these to work out the best projection. We want the distances in the projected image to be as close as possible to the distances in three or more dimensions. The most popular dimension reduction techniques are principal component analysis for continuous data (though it is often used for other types, usually without disastrous distortion of the underlying patterns) and correspondence analysis for categorical data.

They will both provide two-dimensional coordinates for each of the observations, and these coordinates can then be encoded, for example, in a scatter plot. Even without one of these techniques, as long as you can calculate distances, you can feed them through multidimensional scaling.

A major area of theoretical work in statistics in recent years has been sparse data. This is the norm with data on people's social media, internet use or other routinely collected data, and especially affects data that has been linked together from different sources. Perhaps it is known that you and I are both interested in data science. That is one similarity. My wife and I both like Italian food. That's another similarity. Nothing is known about your culinary preferences or my wife's views on data science.

How similar are you to her? This is a distance measure, and if we are to visualize this sort of data, we will need to calculate it for everyone, not just where there are matches, but it can usually only be done with some assumptions. Perhaps the absence of information means that you don't like Italian food and Mrs. Grant can't stand data science. Or perhaps we allow people to partially like things, so we can fill in the average. Whatever you do in the data will not be evident in the visualization, so it needs to be explained.

Projections provide a scatter plot of some abstract combination of variables against another abstract combination of variables, which is really not a good idea for readers' understanding (we encountered something similar in Chapter 11 with complex predictive procedures like neural networks). I have several times had an awkward conversation with a colleague when I showed them a scatter plot arising from principal component analysis only for them to ask me what the horizontal and vertical axes show. It's not an easy question to answer if they don't have any mathematical background.

Biplots and symmetric maps were devised as a way of indicating not only the data but also where the original variables lie in the projection (Figure 12.8). Suppose we have that cube containing dots in three dimensions, and we draw lines along three of the edges radiating out from one corner, for the height, width and depth. If you not only saw the dots in the projected image, but those lines as well, it would be a biplot.

12.5 CLUSTER ANALYSIS

We might be interested in whether our data fall into clusters. In other words, a group of observations with short distances among them, and all the others much farther away. Take another look at the iris flowers' four variables in Figure 6.6.

One of the species is clearly a cluster on its own, while the other two overlap a little and are not so easy to separate visually. However, knowing any two of the variables allows an accurate decision tree to be formed (see Figure 11.2). Even if you didn't know the species, you could still detect these clusters (machine learning people call this unsupervised learning). This cluster analysis is interesting for lots of analysts because it means you can "segment" your data.

You might have a database of previous customers and want to predict who will be interested in your new product. Supposedly personalized internet advertisements and product recommendations are done on this basis. Any projection that preserves as much of the distances as possible is likely to show up the clusters, as the biggest distances are between them.

So, dimension reduction and cluster analysis are closely related. If we identified lots of clusters, we might show them and their connections with other clusters using some kind of network visualization (Chapter 13).

Despite the fascination with clusters, it's important to recognize that the computer, tasked with finding a given number of clusters, will do its best. However, some things just do not occur in clusters, but in a continuum. To take an example from a blog post by Joel Caldwell, our feet are like this: they are variously long, wide, flat. If you have a factory making flip-flops ("thongs" in some parts of the world), you probably only need to make small, medium and large. But if you are making smart dress shoes, you need lots of gradations. Yet, we are classifying the same feet.

In this way, you might have to partition a continuum into different clusters for different purposes. Having a visualization can help to convince a less statistically literate colleague or client to give up on the dream of clusters that just don't exist.

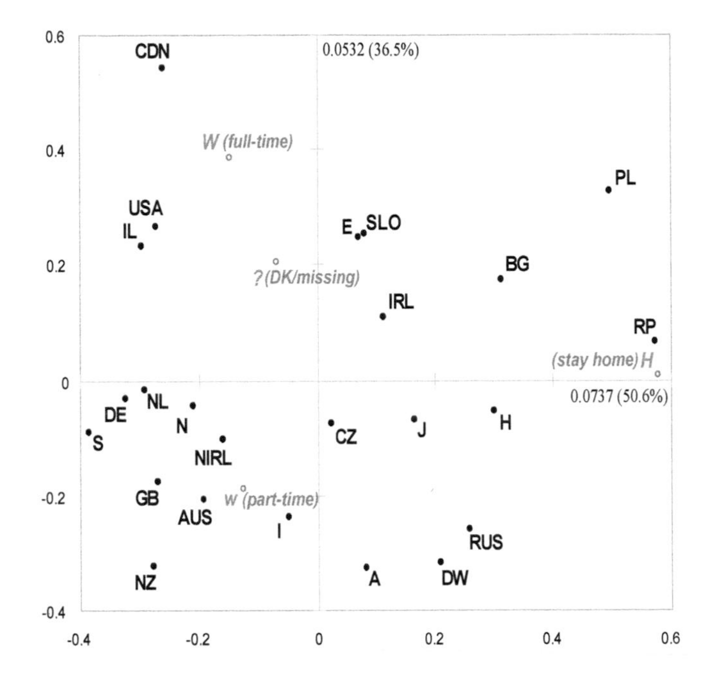

Figure 12.8 A symmetric map summarizing the results of a survey in various countries which asked for opinions on the best role for new mothers to play in the workplace: work full-time, work part-time, stay at home or don't know. The hollow circles indicate the projection of the extreme ends of the data, where a country would be if 100% of participants answered in that category. The projected countries are represented by the dark circles, and the closer they are to the hollow circles, the more preference they have for that category. There is a characteristic horseshoe shape to the country points, which often appears in dimension reduction, arcing from "work full-time" (Canada, United States, Israel) through "don't know" and "work part-time" (Australia, Italy) to "stay at home" (Romania, Bulgaria, Hungary, Poland). The two dimensions can also be interpreted as measuring a work / don't work spectrum (horizontal) and mix / don't mix work and childcare (vertical). From *Correspondence Analysis in Practice* by Michael Greenacre. Copyright CRC Press.

Figure 12.9 The Saturn-shaped data from Figure 12.7 have data in three dimensions (left–right, front–back and top–bottom), summarized here by principal component analysis (left) and t-SNE (right). Principal component analysis retains the two dimensions which give highest distances (variance) among the observations, viewing the planet from above the north pole and ignoring the height of the spherical part of the data. Points at the north pole and the south pole of the planet will appear together. t-SNE tries to show that, even within the sphere, some observations are closer together than others, and introduces distortions to capture this. However, in doing so, it breaks up the structure of the rings. This is because t-SNE is an algorithm that tries several distortions in an effort to find a good one, but needs to be fine-tuned to produce its best results. The rings contain fewer data than the planet, and so they don't matter so much to the algorithm.

A relatively new method to project multivariate data into two dimensions while identifying clusters as much as possible (perhaps at the expense of preserving the distances) is called t-distributed Stochastic Neighbor Embedding (t-SNE). This is like projecting the dots, not onto a flat sheet of paper, but one that curves up and down (a similar idea to support vector machines in Chapter 11). Finding the best way to bend the image is a very difficult task; it can take a computer a long time to do this and it might not find the ideal image, but t-SNE can still be very useful (Figure 12.9).

Dendrograms are a compact way of showing how observations go together into clusters. Imagine we start with a small sphere

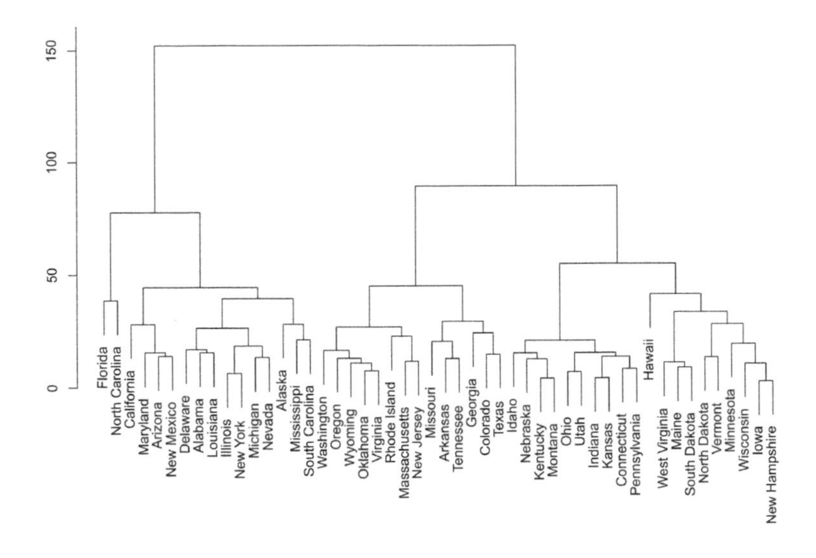

Figure 12.10 A dendrogram uses a tree format but the vertical distances between branching points shows how different they are. In this case, rates of various violent crimes are compared among American states. Starting from the right, we can see that Iowa and New Hampshire are very similar to each other, then Wisconsin is slightly more different, and Minnesota more different still. As we widen the search, we pick up North Dakota and Vermont, which have already formed one cluster together as they are similar to each other. So on up the tree, until we find four and then two super-clusters. The task of cluster analysis is to decide how far up this tree to go before stopping and declaring the clusters to be meaningful.

around each of our dots, and gradually expand them all. When they meet, like bubbles, they join together. As time goes by, our data form into clusters. Now, if we encode the size of those expanding bubbles to the vertical position, and line up the observations along the horizontal position, we can draw lines going upward from each observation, which combine at the point where their bubbles combined (Figure 12.10).

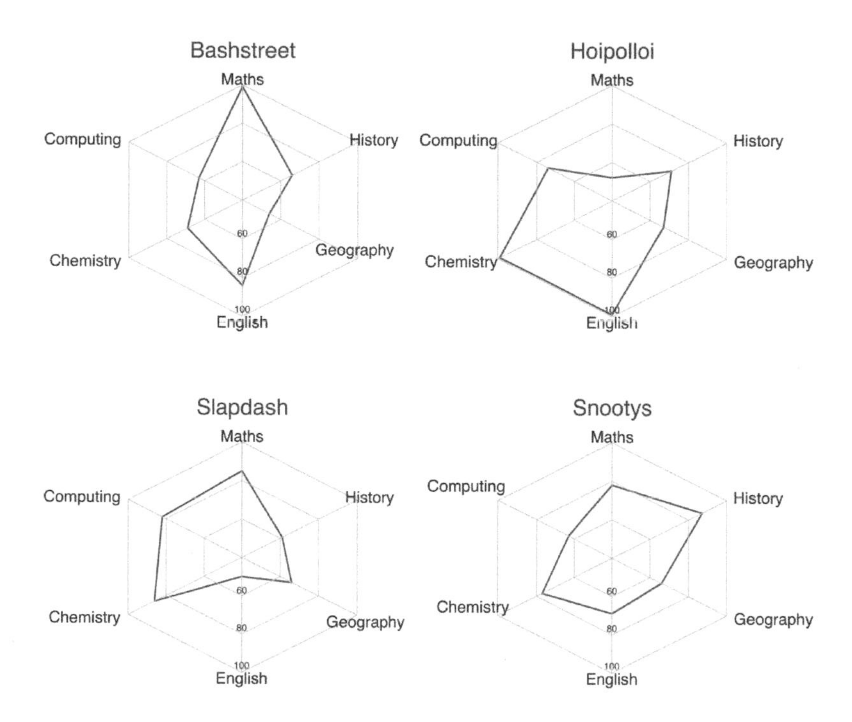

Figure 12.11 A radar chart showing four fictitious schools' marks on six subjects. Each subject is a variable and each school an observation.

12.6 OTHER APPROACHES

There have been many attempts to visualize multivariate data. Radar plots (Figure 12.11) and stars are popular: lines extend out from a central point, with each line having one variable encoded to its length. The resulting shape gives you an indication of the several values that each observation has across its variables. The stars can even be located in a scatter plot, thus including a further two variables in the visualization. The objection I have to these is that it's hard to sense the relative length of lines that point in different angles and start in different places. I'm also not convinced that the overall shape (round, oblong, crescent...) is really perceived consistently by readers.

Line charts comparing matched data, which we encountered in Chapter 3, can be repurposed for multivariate data, with different variables at each horizontal location, rather than time. Combined with smoothing or maybe edge bundling, this could show some clusters quite effectively.

Andrews' plots take a series of wave shapes such as you might recall from trigonometry at school – sines and cosines – and allocates one variable to each. Each one is multiplied by the value of the variable and they are added together to give a complex wave shape, which is the same way that some musical synthesizers work. One might hope that similar observations will have similar-looking waves. It's a nice idea but there is no guarantee that the same change in variable will not produce two very different impacts on the final shape.

Finally, Chernoff's faces is a much-loved method that is very rarely used. A face is drawn for each observation, and variables are encoded to features like the size of the eyes, orientation of the eyebrows, and so on. Like Andrews' plots, it sounds good, and it is certainly fun, but there is no guarantee that the same stimuli won't produce very different impacts on the reader. After all, we put a lot of attention into the shapes of eyes, eyebrows and mouths, which indicate emotion. Not so for the size of people's ears. As William Cleveland wrote of them, "visually decoding the quantitative information is just too difficult."

CHAPTER 13

Maps and networks

MAPS ARE VERY POWERFUL FORMS of data visualization. The reader will usually know where to look for the area that interests them, which makes the data rapidly understood and absorbed. There is a long history of cartography that we should draw on when making maps for data visualization, though most statisticians or data scientists do not know enough about it.

Visualizing network data, where observations are linked in some way with other observations, shares many of the same ideas as mapping geographical data. The network might reflect physical proximity, or social connections, investments and trade, or any other concept that involves connections.

13.1 MAPPING BASICS

Our planet is not two-dimensional, so drawing a map involves some compromise to get a roughly spherical surface onto a flat image. There are several projections that can be used, which are like peeling an orange and laying the peel out flat: it either has to be split in places, or stretched somehow. Without going to a lot of detail, the important considerations for making and reading maps with data are as follows:

- The most familiar projection, the Mercator, is very distorting as it gets away from the equator; Greenland becomes huge and Antarctica stretches right across the bottom of the world map.
- If all the data are contained in a fairly localized area, and the

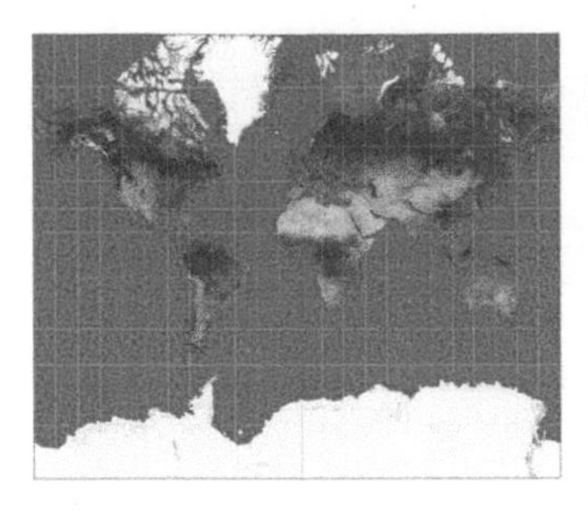

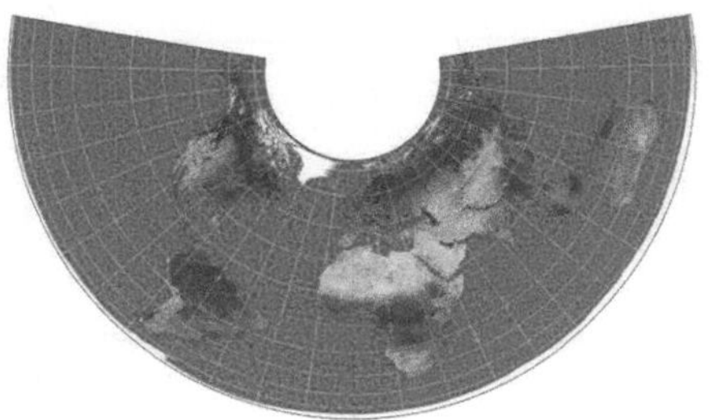

Figure 13.1 Two popular projections for maps: Mercator (left) converts latitude and longitude to vertical and horizontal positions respectively, but this makes places farther from the Equator look bigger, while Albers (right) preserves relative area but at the cost of shape. There is no perfect projection. Images from Wikimedia by user "Strebe," CC BY-SA 3.0 license.

projection is centered there, the choice of projection won't matter much.

- Some projections that avoid distortion unfortunately involve splitting the map.

Typically, when using a map in the data visualization context, we need to decide how much information to have visible in the map. After all, less is more if we want the data to be clear. Maps are comprised of layers: one might just indicate shorelines and whether a place is under water or on land, whereas another might have country boundaries and names, then another cities and towns, another roads, another rivers, and so on.

Sometimes the information in these layers matters to the reader, but if not, then it is probably a good idea to omit them. Just a few landmarks can help readers find their way, such as a coastline or river and a city boundary, but unfortunately this isn't always possible (Figure 13.2) and some readers will not know the area.

As we add data on top of the map, we might find it gets obscured to such an extent that readers can no longer tell where the data are supposed to be located. Semi-transparency can help, but ultimately there comes a point where the data are too dense to

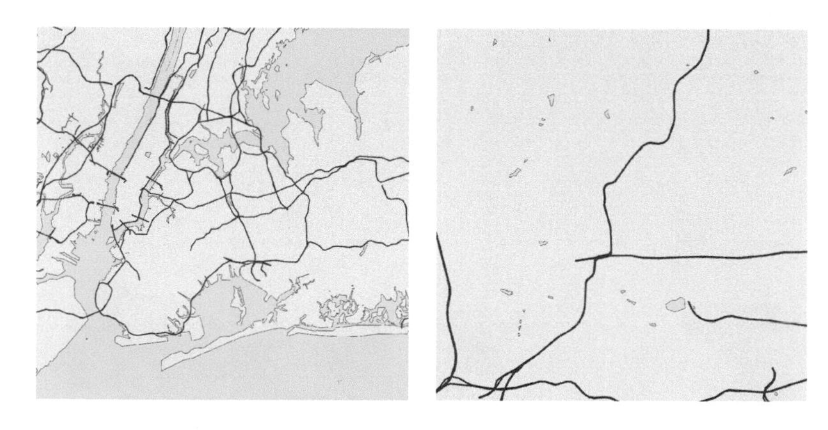

Figure 13.2 A map with just water and major roads is an adequate background to display data for New York (left) but not Johannesburg (right). Copyright OpenStreetMap; prepared with Mapbox.

allow the map to be seen, and they have to be summarized. The next section looks at the options for superimposing data.

A very common encoding is for regional data values to control the color in those regions of the map. This is called a choropleth. The usual limitations of perceiving changes in color apply. Choropleths lend themselves well to small multiples (see Figure 12.1). Because the color blends right into the background, more variables could even be added on top, but we have to be careful not to overload the reader; user testing can help.

One problem that often affects choropleths is that some large rural areas have a disproportionate impact on the reader's first impressions. If the data are about humans, then that could be misleading. Cartograms attempt to correct this by shrinking sparsely populated regions and enlarging densely populated ones. This can be done by replacing the map with a grid: hexagons are best because they minimize the error where the grid has to approximate the boundaries of a region.

Recently, cartogram grids of various shapes have been used not so much to correct for population as simply to have one per region, to make the map simpler and maybe more eye-catching. Not only hexagons but circles and other non-tessellating shapes can be employed. I live in hope of seeing a cartogram in jigsaw puzzle shapes one day.

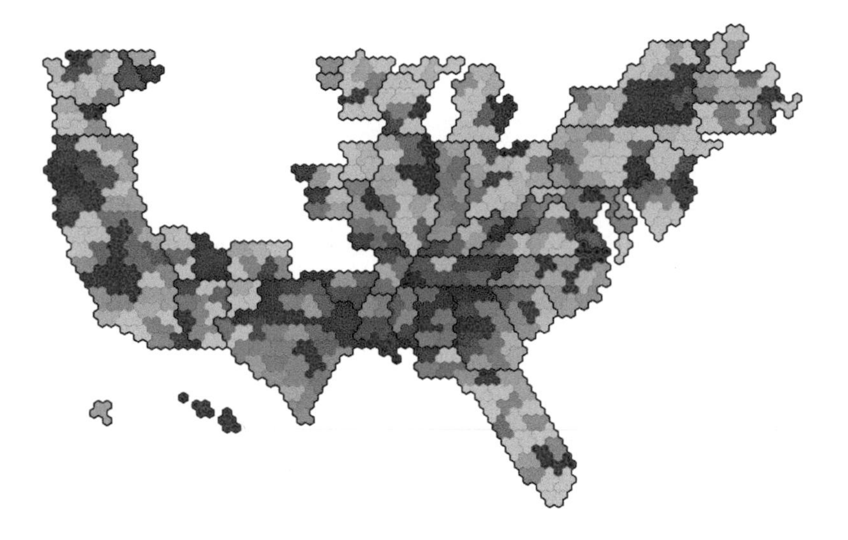

Figure 13.3 American congressional election results in 2012, visualized in a choropleth with hexagonal cartogram boundaries. Share of the vote is encoded to a diverging color scale, with pale gray in the center. The sparsely populated regions in the Rocky Mountains have been pulled apart to preserve as much of the recognizable coastline as possible. This map was created to be viewed online, with interactive information popping up as the pointer hovers over a hexagon. Created by Alec Rajeev, reproduced with permission.

Another approach is to use an algorithm called the Gastner-Newman cartogram to stretch and squash regions smoothly, although this seems to have been fashionable for a brief period before falling out of favor. It is certainly harder for a reader to judge the area of an organically stretched and squashed shape than it is for a regular grid. Also, the familiarity of the map is at risk of being lost if we distort it too much.

Future-proofing is another general concern. One day, your map will be out of date. Perhaps that doesn't matter given the data you are displaying, but one option to consider is linking to an online source for the background image (although some work potentially has to be done to keep the link alive as years go by).

13.2 DATA ON TOP OF MAPS

Apart from the choropleths and cartograms already mentioned, there are many ways of superimposing data that have been tried with varying degrees of success. Markers can appear at specific locations relating to the individual data points (Figure 13.4), or at the centers of regions to show statistics for the region (Figure 13.5), with all the encoding options available to us. Lines can join them to indicate some network (Figure 13.6), journeys or the paths of hurricanes.

A map by itself requires little explanation, but once data are superimposed, readers will probably need labels on the maps, and legends explaining encodings like the color of markers.

We can also superimpose small charts for each region or location of interest. One often sees pie charts (Figure 13.5) and bubble charts (Figure 13.4) superimposed like this, though they are flawed, as previously discussed, and could be effectively replaced with waffles. A simple way of showing a single value at each location is to add rectangles of the same size and fill them with color to a proportion indicating the value. This is preferable to having a symbol at each location, and encoding the value to the size of the symbol, because of the usual problems of perceiving relative areas.

Dot density maps are an alternative to choropleths. They contain regions, with values for each region. Randomly located small markers are drawn within each region at a density (markers per square inch, for example) that encodes the value of interest. This is engaging for readers because it looks like individual data points are drawn on the map, but the drawback is that they might misunderstand and assume that the points really are data. Also, it is

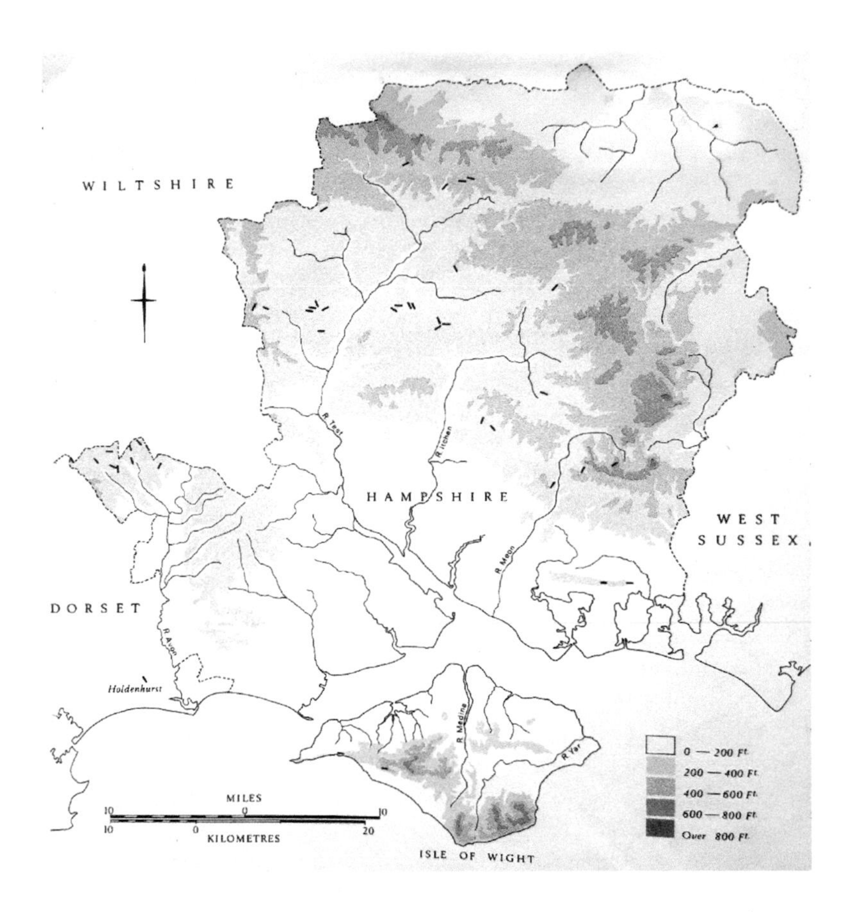

Figure 13.4 Location and orientation of long barrows (elongated mounds of chalk or earth constructed in the late Stone Age) in Hampshire and Isle of Wight, England. In this map, the small oblong markers show both location and orientation without adding any clutter. Only the shoreline, rivers and five categories of altitude are shown, which is just enough to show location and relationship with the landscape. Reproduced with permission of English Heritage from *Long Barrows in Hampshire and the Isle of Wight.*

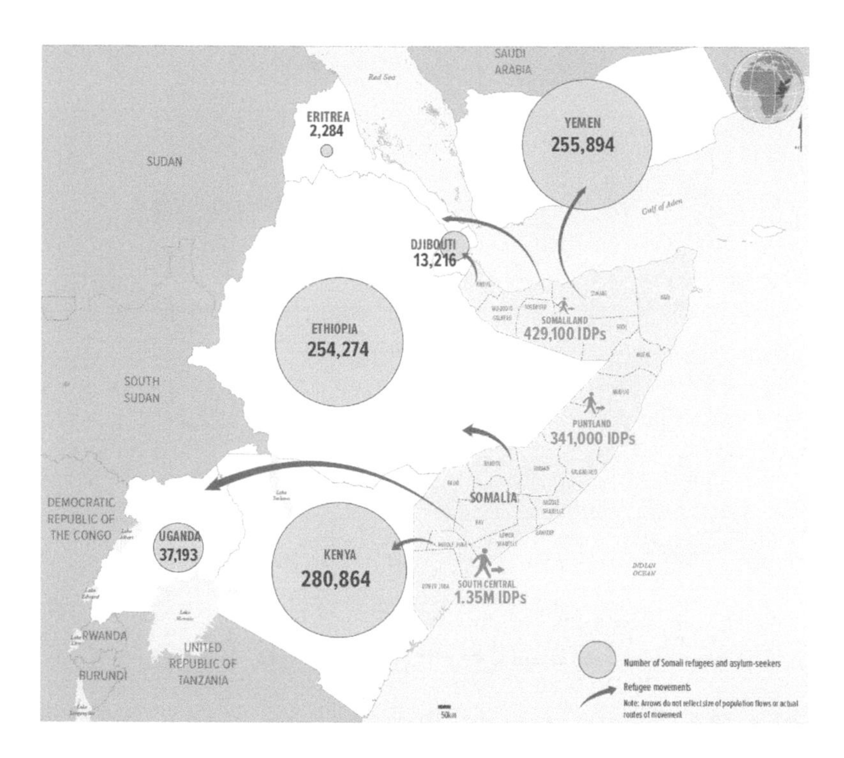

Figure 13.5 Bubbles representing the number of Somali refugees in each neighboring country, superimposed on a map. Reproduced under CC-BY-3.0 license, produced by United Nations High Commissioner for Refugees Operational Data Portal.

Figure 13.6 Semitransparent lines between regions indicate commuting for work from the 2011 UK Census in an interactive map. Copyright Office for National Statistics.

Figure 13.7 Locations of deaths in the 1860 cholera outbreak in Soho, London, shown as hexagonal bins over a map. Background map by OpenStreetMap contributors, data collected at the time by John Snow and made available online by Robin Wilson.

very hard to, as Cleveland said, translate the density back into a number when looking at the map.

Maps with gazetteer symbols are very common: an ear of wheat to indicate agricultural land, a cog to indicate industrial zones, and so on. Although these could in principle be used for data visualization, there are usually too many different symbols for readers to absorb the pattern easily.

If the data reaches overload, we can switch to small multiples. Naomi Robbins has popularized minimap symbols in tables, although this depends on the distinctiveness of the shape that is shown with no other markings on it: good for Texas, less so Colorado (Figure 14.2). Another option is to overlay hexagonal or rectangular bins on the map (Figure 13.7).

We can draw contours, like showing height above sea level (Figure 13.8). This can work well, especially with software that can add shading suggestive of light shining on peaks and troughs. However, it can also clutter up any visualization that is already busy. It also relies on the reader's ability to look at contours and understand quickly what they are being shown.

Animation is an excellent way of including time in a map, for

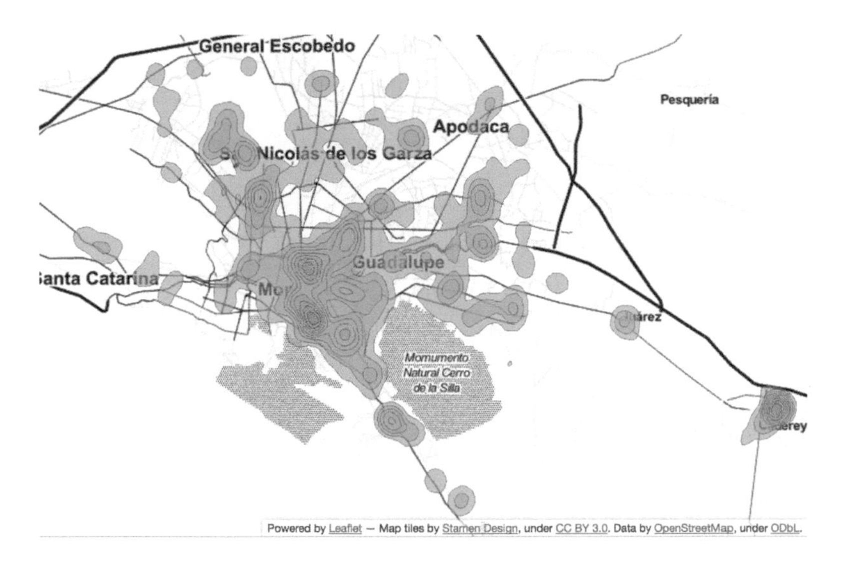

Figure 13.8 A contour plot of the number of homicides in Monterey, by Diego Valle-Jones. Reproduced with permission.

example Chris Whong's video of New York taxis over the course of one day, *NYC Taxis: A Day in the Life*.

Interactive online maps have the added benefit of allowing the reader to zoom in on regions of interest, perhaps to click for more information, and to switch on and off the display of layers in the underlying map. Interactivity is explored in Chapter 14.

Geographical location can also serve as a way to help readers absorb information quickly, even without having any visible map. Figure 13.9 uses a cartogram layout of the United States to arrange small multiple line charts.

13.3 SPATIAL MODELS AND UNCERTAINTY

Uncertainty is a matter of life and death when showing predictions of hurricane paths, and there are at least three ways of showing this that have become quite common, shown in Figure 8.4. Another approach, if it is possible to obtain repeated statistics from the bootstrap (see Chapter 8) or a Bayesian model, is to produce an isarithmic map, which is like kernel density in two dimensions, with the density encoded as color in the manner of heatmaps (this is the approach in Figure 13.8).

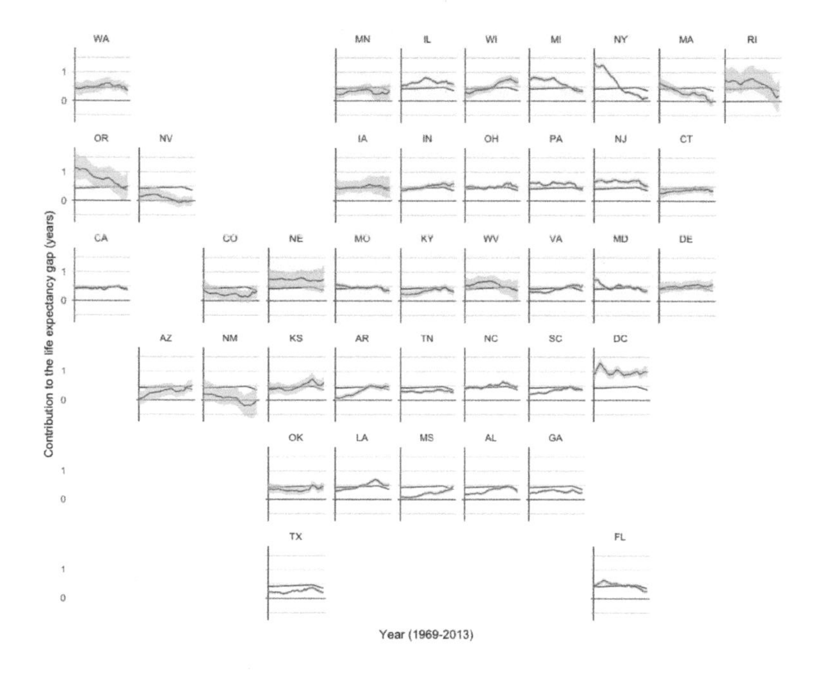

Figure 13.9 A small multiple grid map showing smoothed trends (1969-2013) in the contribution non-communicable diseases make to the gap between black and white American men's life expectancy. The black horizontal line indicates no gap. Several features make this easy for readers to digest. Placing the individual states' charts close to their geographical location means they can be found quickly. Including the national trend as a repeated red line in each chart allows immediate comparison. Smoothing and drawing not just the line but the confidence interval gives an impression of trend including uncertainty. Reproduced with permission from "Long-term trends in the contribution of major causes of death to the black-white life expectancy gap by US state" by Corinne Riddell and colleagues, reproduced with permission.

Spatial models extend the regression techniques from prior chapters to include spatial autocorrelation (see Chapter 9 for autocorrelation in time). The goal is to find a smooth underlying density that might have resulted in the isolated observations that we have in our data. It's possible to adjust for some confounders or time effects in the model too. Often, Bayesian software is used for this sort of model, which provides simulated values at different points over the map, and allows us to easily draw an isarithmic map as output. An increasingly popular technique for this is called Gaussian processes.

13.4 NETWORKS

Networks are just data with additional information on connections between observations or clusters of observations. It can be physical connections, a social network, or some measure of proximity. Visualizations of networks can get very tangled and hard to understand, so there are some tricks to simplify them.

If we reduce the distance matrix (which appeared in Chapter 12) to a binary value – connected or not connected – we can start to draw a network of connections. This is called an adjacency matrix. Typically, there is a marker for each observation and a line between them for each connection. Unless there is some reason for encoding some other variables to the positions of the markers, they can be placed anywhere.

Deciding where to place the points for clarity is not easy, but there are algorithms to calculate the best layout. The connections can also be tidied up by software that uses edge bundling, which pulls the lines together if they are starting and ending in similar places (Figure 13.10). This makes it much easier for the reader to absorb.

Networks simply show an adjacency matrix, which is not much information in statistical terms. The markers could have other variables encoded as size, shape, color, etc., though we would run the risk of overloading the reader with too much information all in one visualization.

The connecting lines are sometimes thickened or made into several parallel lines to indicate something about the strength of the connection, though with edge bundling it would be very hard to make out what is a thick/multiple connection and what is a bundle of connections. It is hard to imagine how uncertainty could be

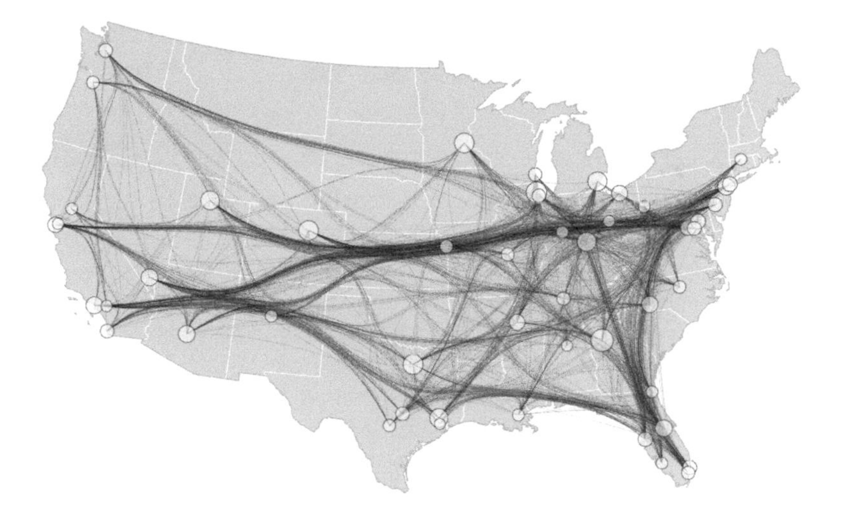

Figure 13.10 A network showing airline routes within the United States, using edge bundling and semi-transparency to clarify the patterns and avoid the spaghetti effect. For example, flights from San Diego to Houston initially curve north so that they can join with other flights from the west coast; this is not what happens in reality. By Sophie Engle, reproduced with permission and under GPL-3 license.

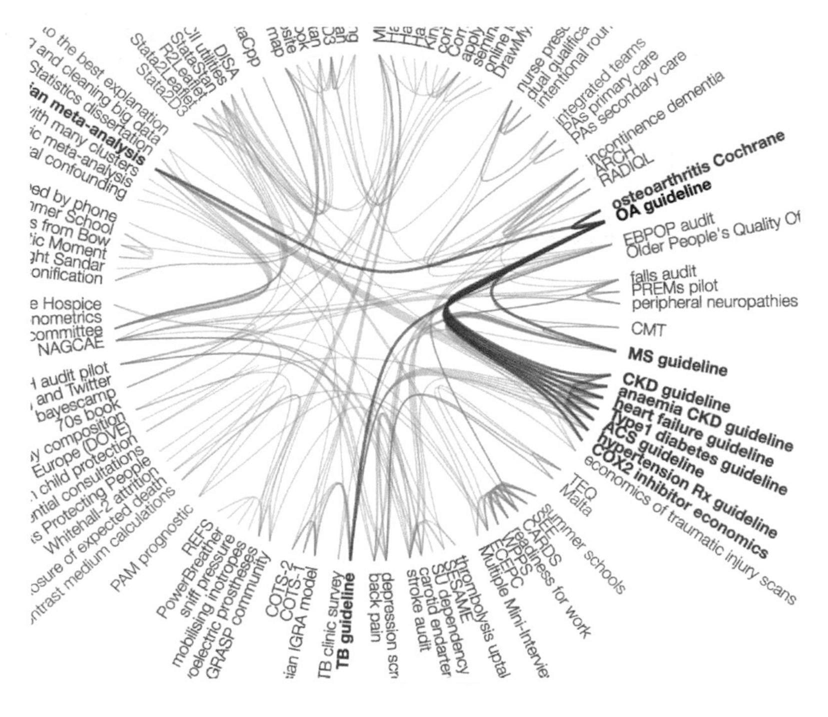

Figure 13.11 Connections between projects I have worked on, visualized as a radial network with edge bundling.

shown about the markers or their connections on top of all the information already present, without the use of interactivity.

Circular layouts can reduce clutter. The marker can be omitted entirely because the labels will be unambiguous, and the location around the circle can be tweaked to get clarity through edge bundling (Figure 13.11).

Networks are sometimes subjected to creative formats, such as making them look like subway maps; be careful not to let your format get in the way of easy reading of the data. The job of the visualization is to convey quantitative information easily to the reader, not to show off design skills.

CHAPTER 14

Interactivity

THE RECENT RAPID GROWTH in data visualization has been in large part driven by the power of interactive content. Data visualizations can be shown in a web browser and then take advantage of the interactivity the browser provides, for example showing more information when the user hovers the mouse over a marker, or clicks on a region in a map. This interactivity used to rely on proprietary software like Adobe's Flash, but is now achievable by everyone through JavaScript, the programming language that powers web pages. This chapter outlines various forms that interactivity takes, with links to good examples. However, as links occasionally break and online content is lost, you will find the most up-to-date links at this book's accompanying web page: **robertgrantstats.co.uk/dataviz-book**

14.1 WEB PAGES AND JAVASCRIPT

A web page is simply a plain text file that contains instructions on what to display where. These are written in Hypertext Markup Language (HTML), and can also bring in formatting instructions in a language called Cascading Style Sheets (CSS), and most relevant to data visualization, a more general-purpose programming language called JavaScript. Confusingly, JavaScript has nothing to do with Java, which is a different language for making user interfaces (you may have encountered Java applets, which have been popular for teaching statistics).

With JavaScript, we can supply instructions to the web page like "when the user clicks on Texas in the map, display a table of statistics about Texas underneath it." We don't have to write everything from scratch because there are some useful functions that we can link to and then use. The most flexible and widely used "libraries" of these functions are probably D3 ("Data-Driven Documents") for having data determine – and change – anything on the page, and Leaflet for mapping, but you should check for any interesting newcomers.

For data visualization, we are really leveraging existing technology intended to make web pages engaging and useful, and over the last twenty years huge investments have been made in browsers, so however complex a task we might write in JavaScript, it should run very quickly in a modern browser.

Not every browser has the same capabilities. They are engaged in something of an arms race, so if you are making an interactive visualization, you should check that it works as intended in each of these. If you are viewing one that someone else made recently and it doesn't seem to work, try updating your browser.

Although we can make content to be viewed in the browser, it doesn't have to be online and publicly visible. If confidentiality is a concern, it can be viewed locally on one computer or through a protected company server. Increasingly, online web pages will be viewed on tablets and smartphones.

Designing a web page so that its contents are adjusted to be most legible depending on the screen size, internet bandwidth and other factors is called responsiveness, and is a major concern of web developers worldwide. There are JavaScript libraries which can make this task a lot easier, but moving text around is a far simpler task than moving elements of a visualization around to fit the screen, so be prepared for another layer of sketching, experimentation and user testing if you want to go responsive.

14.2 FORMS OF INTERACTIVITY

There are many different ways that a reader could interact with a visualization and attendant tables and text. They are more likely to do so if it is obvious what to do – if it is familiar. These are approaches used fairly widely at present:

- Hover the mouse to bring up a tooltip (floating box with additional information).
- Click to show additional information nearby.

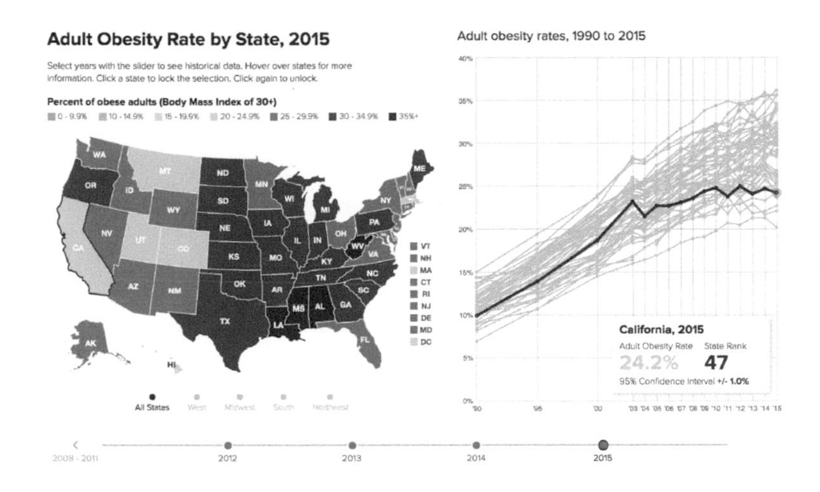

Figure 14.1 Interactivity in a map and linked line chart in the online State of Obesity annual report

- Click to replace content with more detail on the object you clicked on ("drilling down").
- Click and drag to zoom in on the selected rectangle.
- Pinch or widen fingers on a touchscreen to zoom in.
- Scroll or swipe down to move to the next level of detail ("scrollytelling").
- Move sliders to control some aspect of the visualization.
- Toggle between showing and hiding different aspects.
- Animate the content unless paused by the user, possibly as an introduction to the interactive content that follows, or to show different possible outcomes, bootstrap samples, etc.
- Move through the story as arrow buttons are pressed.
- Scroll around and zoom in on a map (also, we can roll a virtual globe around).

The page might contain more than one connected visualization. A good example is the State of Obesity annual report (stateofobesity.org, Figure 14.1). Let's look at the connections from the page

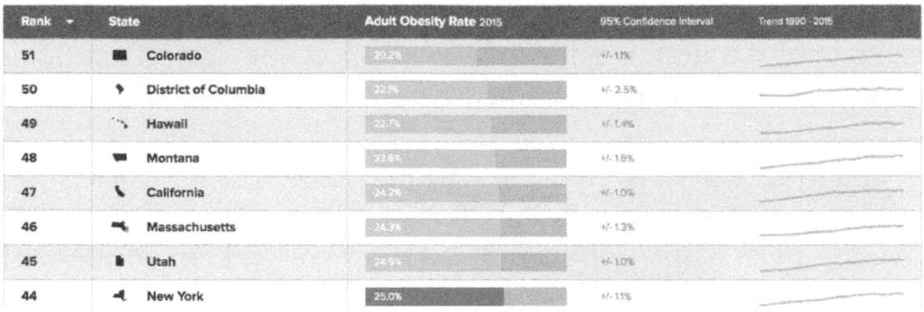

Rank	State	Adult Obesity Rate 2015	95% Confidence Interval	Trend 1990 - 2015
51	Colorado	20.2%	+/- 1.1%	
50	District of Columbia	22.5%	+/- 2.5%	
49	Hawaii	22.7%	+/- 1.4%	
48	Montana	23.6%	+/- 1.6%	
47	California	24.2%	+/- 1.0%	
46	Massachusetts	24.3%	+/- 1.3%	
45	Utah	24.5%	+/- 1.0%	
44	New York	25.0%	+/- 1.1%	

Figure 14.2 Alternative presentation by bars and sparklines in the State of Obesity website

on statistics on adult obesity. First, we see a choropleth map of United States states on the left, color coded by obesity prevalence, and on the right a line chart with one line per state, obesity on the vertical and time on the horizontal. All the lines are gray.

There's a slider at the bottom that lets us turn the clock back in the map, and the color coding changes accordingly. When we hover over a state, the relevant line is highlighted in black in the chart, with a small blue circle indicating the year that the map currently shows – subtle and sparing highlighting gets the message across while avoiding information overload for readers.

The same highlighting happens from hovering on the line chart: the relevant state gets a black border in the map. This way we can look at the best and worst in the line chart and see where they are. If we click on the state or the line, the line and state stay highlighted and we can hover along the line to see statistics in different years. When we scroll down, we find the next layer of detail in a table of states (Figure 14.2). This presents the information with mini maps, bars and sparklines. If we click on the state name there, we drill down further into a state-specific page, which breaks down the obesity prevalence by age, race and sex.

The layout has changed in recent years, but there also used to be policies and laws listed on this page, so you could see a list of policies that could help or hinder obesity prevention, with green tick or red cross icons as to whether they had been implemented in that state. Clicking on the policy then took you through to an in-depth political analysis. So, each user could drill down to the level of detail they wanted, from headlines to deep discussion.

To see "scrollytelling" in action, an excellent example is a *Guardian* online article called "Bussed Out" (goo.gl/uDRgKQ).

For an example of how interactivity allows the layering of huge amounts of detail, an innovative example is the *New York Times*' "How the Recession Reshaped the Economy, in 255 Charts" (nyti.ms/2jVJvTM).

Older interactives used software called Flash, rather than JavaScript. This allowed some of the interactivity we've just seen: hovering over a line chart and having text pop up in a tooltip, and scrolling the timeline back and forth. The technology of interactivity doesn't matter as much as careful thought about what you are going to do with it. My favorite visualization remains one made by Amanda Cox for the *New York Times* in 2010 with Flash (goo.gl/RtUPB2).

It has a clear and simple design and presents the story in six layers of detail. I'm sure this drew in lots of readers who would have skipped it if it had been laden with detail at first glance. It shows each year's American federal budget forecasts as pale blue lines extending upward, departing from the reality that unfolded (a thicker, dark blue line). Because the forecasts were always optimistic and pointed up from the real dark blue curve, Cox called it the porcupine chart.

The reader might think it through like this:

1. The US government's debt is coming down.

2. Back in 2009, it got into a lot of debt.

3. It seems to have bumped up and down over the 30 years before that.

4. But every single year, their forecasters said things would get better.

5. That forecast depends on a lot of assumptions, like how many people will be unemployed.

Highlighting of specific years is done by a small, semi-transparent, yellow circle. Because there is no other yellow on the page, and it contrasts with blue, it draws the reader's attention with very little clutter added to the image. Rather than assume the reader understands it in the same way as you, user testing can really help to refine these complex visualizations.

14.3 METHODVIZ

The TensorFlow Playground (see Section 13.3) is a beacon of clear visual explanation in the world of machine learning methods. Visualization does not usually make much inroads in this area, but here we see an interesting new angle: not dataviz but methodviz, showing how the method works and helping people understand and use it effectively. The best examples of this are interactive to some extent.

Statistician and medical researcher Paul Lambert made a page that shows how splines fit data, given some controlling parameters (le.ac.uk/hs/pl4/spline_eg/spline_eg.html). Google's Ian Johnson, Martin Wattenberg and Fernanda Viégas made a page about the t-SNE method for visualizing clusters in multivariate data (distill.pub/2016/misread-tsne/), which we encountered in Chapter 12.

14.4 RUNNING THE ANALYSIS TOO

Expanding the idea of presenting different layers of detail, and different questions of the data, for different readers, we might reach a point where there are so many potential angles to accommodate that it would be easier to upload the data and let the reader choose for themselves. In fact, it is possible to program quite an open-ended set of analytical and graphical tools in JavaScript and give the reader this freedom. It will be a much harder task than making a more closely curated interactive page, but it could engage readers more deeply in understanding the data.

Bayesian data scientist Rasmus Bååth made a page that not only explains a Bayesian method to compare means between two groups of unmatched data, but also runs it without requiring any software. It is possible to replace the data provided with your own data and run the comparison (sumsar.net/best_online/).

The Locks, a family of statistics teachers, made a website called StatKey that performs a variety of analyses using computer simulations and randomization methods (lock5stat.com/statkey). Again, it is possible to replace the data provided with your own data. Both these websites use only JavaScript, so they can run on smartphones and tablets too.

These two examples run on the reader's computer or smartphone, which is called client-side. An alternative approach is to have some analytical software installed on a server, which then supplies results as requested from the reader's browser. This is called server-side, and although it requires control over the server, and expertise in programming, there are some simple tools to achieve it, like the R package Shiny.

14.5 SECURITY AND CONFIDENTIALITY

Any client-side online visualization will require an upload of data to a web server. That could be a problem for confidentiality, as it is effectively published and can be downloaded by visitors who know a little about websites. The data might also move to a different legal jurisdiction, depending on where the internet service provider's servers are physically located. Anyone making a visualization like this with sensitive data needs to think through the implications before they start.

Because the JavaScript program to run the interactive visualization has to be downloaded and run by the browser, the reader can open that program and read it. This means that the program is essentially published and it is sometimes possible to learn from and adapt other people's JavaScript work (of course, one should give credit and not just copy without permission).

When D3 (which is client-side) was relatively new, everyone was learning to use it, and it was possible to view the source code behind even the most highly regarded sites like the *New York Times*, and understand it. This was a major driver of the explosion of clever visualizations with D3 in the first half of the 2010s. But as time has gone by, commercial pressures have pushed the programs into large, all-purpose in-house JavaScript files, without human-readable annotations. It is also possible to use software to "uglify" the JavaScript (deliberately obscure the meaning of the program).

CHAPTER 15

Big data

WHAT HAPPENS WHEN THE DATA GETS SO BIG that calculation becomes impossible? Big data is a term that attracts a lot of hype, but it is real. What if the data is so big, it can't just be stored in one file and opened by your favorite software? The most obvious meaning of big data is that the data is bigger than your memory chips.

Another meaning is that it is arriving through some kind of stream and has to be processed and visualizations updated in real-time. There are some other definitions that have been proposed, often a series of adjectives starting with the letter V, but these two (volume and velocity) are the only ones I recognize as problems of big data. The others, like veracity, are problems of any data.

Apart from the technical challenge of working with the data itself, visualization in big data is different because showing the individual observations is just not an option. But visualization is essential here: for analysis to work well, we have to be assured that patterns and errors in the data have been spotted and understood. That is only possible by visualization with big data, because nobody can look over the data in a table or spreadsheet.

The important considerations that we've already encountered in this book, such as showing uncertainty or periodic effects, are still important in big data. There is sometimes a notion that if the data are "big," then all those problems disappear. Sadly, this is just not true. A billion biased observations are just as biased as ten.

A nice example dataset, not quite big enough to challenge serious computer hardware, but big enough to learn from, is every taxi journey taken in New York in 2013. There are geographic and time series variables, as well as continuous and categorical ones. This is available online thanks to mapmaker Chris Whong, who obtained it under Freedom of Information legislation (chriswhong.com/open-data/foil_nyc_taxi/). It is about 20 gigabytes when compressed, so choose a good time and connection before downloading.

15.1 TOO BIG

Let's assume we can access the data in some way, not by opening one file, but perhaps a series of large files, or by sending a query to a database server. The details of that are not relevant to the visualization that follows. We can read the data in chunks of a manageable size and add each to a visualization. But if we draw the data in any way like a scatter plot or markers on a map, we will simply have a solid mass of ink on the page (or color on the screen), with no patterns visible.

We have to visualize some statistics instead, but how are we to calculate them if we can't open the data? This is the crux of big data and why it is a genuine area where a lot of clever people are concentrating their efforts. New software and cloud computing options are emerging all the time. However, the general options are:

1. Work out what variables you need to show, define some categories for those variables, and then count how many observations fall into each category. Now, you only have to work with the combinations of categories, not every individual observation.

2. With some statistical expertise, find stepping stones you need for your calculations, that can be added across chunks of the data to give you the results. Get those stepping stone values for each chunk of the data, and add them together. Then, do the final part of the calculation on the stepping stones, not the individual observations.

3. Take a random sample from the data and do traditional statistical analysis on that.

The histogram is the simplest example of Option 1. If we have a trillion observations from a continuous variable, we can still draw a histogram. We just choose where to cut the variable up into bins, open a manageable chunk of the data and count the observations

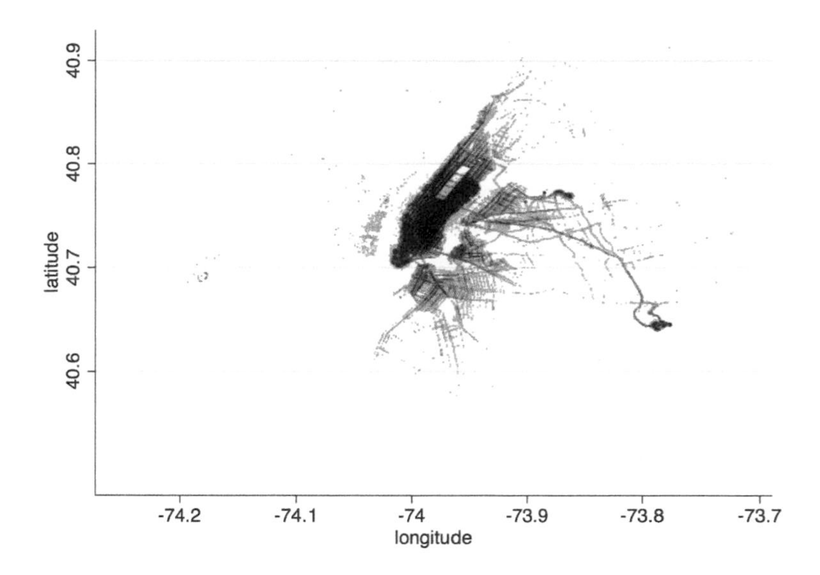

Figure 15.1 A 1000 by 1000 heatmap counting all the starting points of the 179 million yellow taxi journeys made in New York City in 2013. Darker pixels indicate more taxi journeys. It's worth remembering that a simple random sample from the data, of sufficient size, would lead to almost exactly the same image.

falling into each bin. Then we open the next chunk and keep adding to the counts. In the end, we will have a total count for each bin and can draw the histogram, without ever having looked at all the data at one time.

This extends to more than one variable because we can count observations as they fall into combinations of the categories. That is how I made a heatmap of the 179 million taxi journeys in the New York 2013 data; each pixel in Figure 15.1 is a combination of a longitude and a latitude category, with number of journeys encoded to color between black and white.

In this image, I wanted to see fine details of roads, so I made a 1000 by 1000 rectangular grid, but to see the general pattern in two dimensions, a chunkier hexagonal grid, maybe 50 by 50, could provide a two-dimensional histogram.

It's worth bearing in mind that binning data in small rectangular regions allows you to aggregate the counts later into bigger rectangles without having to return to the big data. This applies to more dimensions (variables) too, because cutting each variable into bins results in cubes or hypercubes that can be combined to make bigger cubes or hypercubes.

It also allows some smoothing as you aggregate into bigger bins. As this aggregation happens, some small bins will be next to the boundary between two big ones. If you allocate part of their count to one big bin, and part to the other, you will have smoothed the boundary.

A very widely applicable framework for choosing an approach to big data visualization is "bin-summarise-smooth," proposed by Hadley Wickham: count up data in some sort of bins, summarize them into statistics, then perhaps smooth the statistics. This will get you a lot of exploratory images but of course any analysis that has to crunch all the data will still have to be run somehow, perhaps by random sampling (Option 3).

Kernel density plots are also amenable to the binning and adding approach, because they just add together the heights of the kernel shapes at various values of the variable, but as the data get more plentiful, the benefit of a smooth kernel vanishes along with bumps and gaps that are just the result of noise. Counting and averaging in bins is also useful to show changes: this is the approach taken in Figure 9.2 to reduce the spaghetti of 200 countries over time into typical trajectories, and this can easily be scaled up to big data.

New software using Option 2 is developing all the time. Unless you are an expert statistician, you will not want to derive your own bespoke method. However, it may help to think through a simple example. A linear regression line with one predictor (Chapter 10) can be calculated using

- the number of observations,
- the sum of all the predictor values,
- the sum of the squared predictor values,
- the sum of all the outcome values, and
- the sum of all the predictor values multiplied by the corresponding outcome values.

These are the stepping stones. As before, we can work through the data in chunks and accumulate these five values, then put them together to calculate the regression line, and visualize it. We can do that for one hundred observations, and just as easily for one hundred billion observations – it will just take much longer. We may not be able to visualize every data point around the line, but we can still calculate predicted values and residuals (see Chapter 10) for each observation. Histograms, kernel densities, hexagonal bins, and the like are all good options. All these ways of thinking about big data can be adapted and combined for a wide variety of models.

Option 3 is more controversial. It is simple and very effective because the early years of statistical theory gave us lots of useful methods based on random samples. When we need quick visualizations to find problems in our data, like scatter plots and histograms, and those are for our purposes and not for wider distribution, a random sample will work fine. But, if we are looking for errors or rare patterns, then we must process all the data or we might miss some.

However confident a statistician may be in the power of a random sample, the person asking for the final visualization might object to the idea of throwing away data. Sometimes, the objective is not so much discovery as showing off how big the data are. They might want to tell their boss, "this image contains all one hundred billion observations," and why not? Visualizations serve many purposes.

An alternative use of random sampling for visualization is to show the results of the analysis, for example the regression line, and to give some indication of data by showing a random sample of the data points (this should always make it clear that it's a sample).

15.2 TOO FAST

Imagine the task faced by a stock exchange to track transactions and prices. Many of these are computer-to-computer transactions that happen as fast as possible, to take advantage of fluctuations in price. All this has to be processed and supplied to customers who are paying for a live feed of information. That feed will typically include summary statistics as well as visualizations.

Many organizations nowadays have to operate with live feeds of data like this, not just in finance. Figure 15.2 is an example with

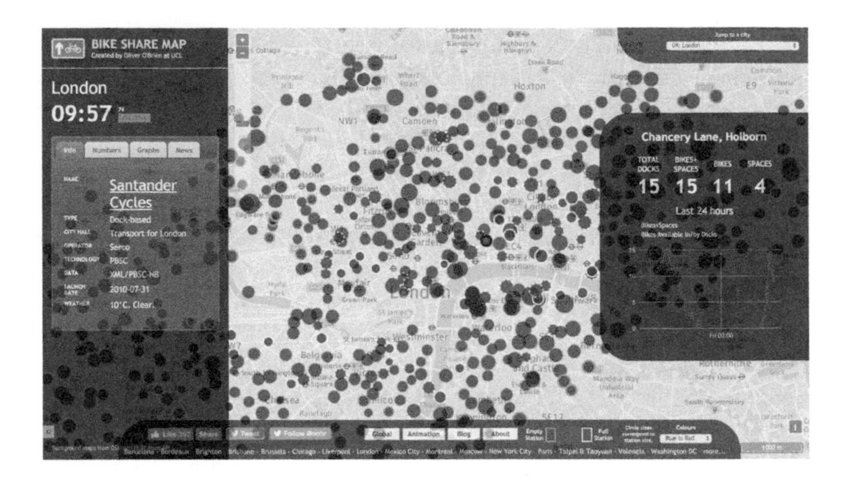

Figure 15.2 Visualization of live data streams on cycle hire in London, by Oliver O'Brien (bikes.oobrien.com/london/). It is 9:57 in the morning, and hire stations on the outside of the area served tend to be empty (blue), while those in central business districts are full (red). By clicking on one (Chancery Lane), we can view more detail over time in a panel on the right. Data from Transport for London, used with permission.

live cycle hire data. Websites dealing with similar challenges on even larger scales include lightningmaps.org or flightradar24.com.

This has to be achieved using some of the same tools as above. The data can provide accumulating statistics and stepping stones for further calculations. The difference is that visualization will typically show a moving average: the stock exchange might have an average price over the last ten seconds, so as new data are added to the calculation, the old ones have to be removed.

Summary statistics, and visualizations, can be based on periods of time, a concept called windowing, and these windows combined in various ways to give the required statistics. It's important to have summaries that include enough history so that they are reliable and informative, not just fluctuating rapidly.

Options 1 and 2 above can be used if you are creating something like a histogram or kernel density: keep track of each batch of counts or kernel heights in the moving calculation, then subtract the oldest batch as a new one gets added. The result can then be provided through a web browser but latency (the time delay between the event happening and it being displayed) has to be minimized so that it updates nearly in real time.

Although real-time updates might sound like a good goal for streaming visualizations, it might not matter to the reader. The technical challenges of getting close to it may not be justified by the way the information is used. Here, as with using all the data in the previous section, the desire to be technically impressive might really matter to the organization or individual, but they should at least be aware of the options and weigh cost against benefit.

Creating a system to feed streaming data through analysis and on to end users is a considerable task and will require a range of programming skills in the team. Fortunately, there are several widely used and trusted open source tools, and with a lot of investment taking place, the options are likely to grow and get ever more accessible.

CHAPTER 16

Visualization as part of a bigger package

A GREAT VISUALIZATION DESERVES TO reach its audience in a great package. That might be a printed document, a website, a dashboard or a presentation. Whatever the package is, it should not be an afterthought to the data analysis and visualization. It should be discussed and user-tested from the outset, to make sure that it will have the desired impact and support understanding of the data.

This chapter will look at some design ideas that are generally not the sort of thing data people consider, but can strengthen the impact of their work. Training in data science or statistics does not introduce design thinking, and there is a misconception in many data people that good design is a thing, not a process, typified by minimalism and Helvetica fonts. Although that is a good starting point, there is no one look that is right, and a process of experimentation, user testing and refinement has to be undertaken. This chapter is mostly for people making visualizations, or paying someone to do it.

16.1 THINK ABOUT IT

We encountered the idea of storytelling linked to data visualization in Chapter 14. When you have a message to convey, think about it as a story. If you are accustomed to writing in a scientific format – aims, methods, results, conclusions – resist the urge to do that in this context. It may well be better to reverse it or at least put the conclusions first.

Another structure you can consider comes from Simon Sinek: instead of telling people what you found, how you found it, then why it matters, reverse that to why, how, what. Try to reduce your message to an "elevator pitch," as short as possible, preferably only a few seconds long. You will have to think carefully about what terms to use and what order to put the facts in. Then use that to structure your visualizations.

However, remember that different readers want different things from your visualization and the package it arrives in. Your story should be the opening for them to explore further. We can layer the information so readers can drill down to their preferred level. What might those layers be? What order will readers want to encounter? What words or symbols would they understand as signposts to different layers?

Dataviz designer Andy Kirk suggests asking:

- What is the reason for its existence?
- Who are you creating it for, and how well-defined are their requirements?
- What function is the package seeking to fulfill?
- What is the likely tone of the design?

The final point in that list is about matching the requirements of the package. Is the visualization intended to be sober, eye-catching, shocking, beautiful, funny? How can you stick with the same encoding but tweak the format to achieve that tone? What could we mean by beauty in data visualization? Noah Iliinsky suggests a beautiful visualization ought to be novel, informative, efficient and aesthetic. Novelty is highly valued in a designer's portfolio, but less so for people approaching data visualization from a scientific background.

A team making a package around visualizations might include a designer, a data scientist, an expert in the topic (like medicine, or transport planning), a web developer, and a manager, and they will have to work together efficiently up to the point of publication, despite the different cultures that they come from. The first step towards this is to understand each other's motivations and to make sure there is some aspect of the project that each team member can be proud of afterwards.

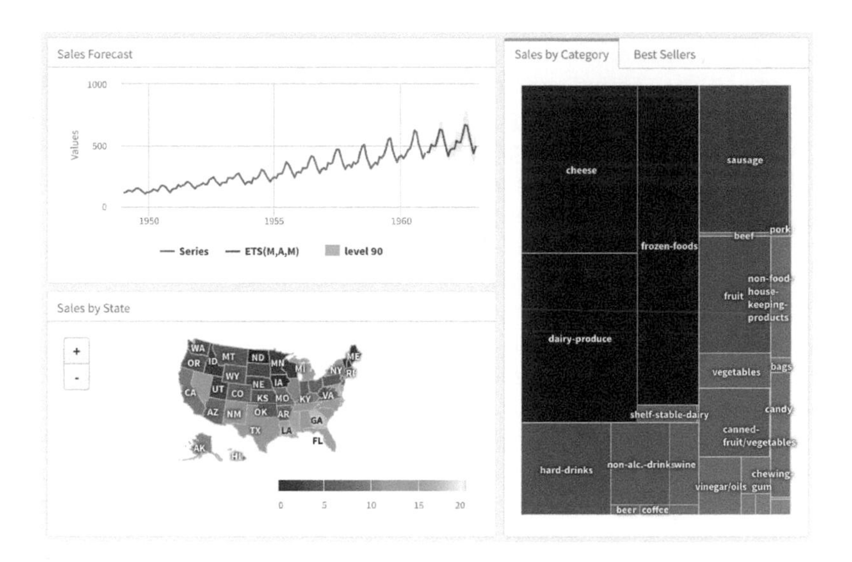

Figure 16.1 Example of a dashboard tracking retail sales data, created by Joshua Kunst and published by RStudio. Reproduced with permission.

16.2 MAKE IT

Dashboards are collections of several linked visualizations all in one place. The idea is very popular as part of business intelligence: having current data on activity summarized and presented all in one place. One danger of cramming a lot of disparate information into one place is that you will quickly hit information overload. Interactivity and small multiples are definitely worth considering as ways of simplifying the information a reader has to digest in a dashboard. As with so many other visualizations, layering the detail for different readers is valuable.

Dashboards are meant to be updated frequently with new data. If you have computer-intensive models like neural networks, they might take too long to fit to the new data (while the boss is waiting anxiously for the dashboard). But you can generate and visualize predictions for new data quickly from previous models, even very complex ones. Some of the problems we have encountered, like prediction out of sample, need to be detected and appropriate warnings added to the dashboard.

The potential problem for most dashboards is that the reader's eyes will dart about trying to find where to start, and if they find this confusing or fatiguing, their first impression will be a bad one. As the curator of information, the dashboard maker should guide them. This could be done by large text headlines against each part, or arrows leading from one visualization to another, or by a separate "How to Read This Dashboard" guide. If the dashboard is interactive (see Chapter 15), the reader could be guided through one visualization at a time with the others faded into the background.

Each visualization needs to be linked visually to the rest of the package. There should be recognizable elements leading attention in and out from the data. If there are pre-attentive cues (see Chapter 7) in text, they should match those in the visualization. For example, highlighted points on a chart can use a color that matches a piece of text alongside.

Some formats of visualization seem to have become standard dashboard components in particular industries. Gauges, which show single values and look like speedometers or dials on some vintage scientific instrument, are one such. They may be fun but are quite poor in terms of the reader translating them back to numbers and comparing one to another.

Color, typefaces, and other styling like line patterns and marker shapes should be consistent throughout. This helps the reader to focus on the information you want them to. Visualizations will probably have to be edited manually to match the design of the report, website, or whatever, and the SVG graphics format is very useful for this.

If it is being made for an organization that already has a logo, color scheme or branding guidelines, you should match that. Perhaps you can identify one color in a logo. Then, there are many websites that will take one color and suggest others that complement it, for example colorhexa.com, 0to255.com, and tools.medialab.sciences-po.fr/iwanthue/. When I made the boxplot in Figure 5.4 and the line chart in Figure 16.3, I picked a color from the organization's logo and generated a scale around that. The choice for data visualization must consider readers with various forms of color blindness, and these websites can help by showing a simulation of how the color palette will appear to them.

Now, with these ideas, let's revisit the train delay data. We have already seen some visualizations of different aspects of the data and different levels of detail. Figure 5.7 (left) gave a quick overview by using splines to smooth the line chart, Figure 2.6 focused on which time of year was worst in each year, and Figure 2.2 is the original, detailed line chart. I combined them into a poster or flyer in Figure 16.2.

I want to take a moment to go through the thought process and decisions in detail. We need to guide the reader from the top level of detail down, so they don't get confused and they can drop out at whatever point suits them. I tried to do this by placing the images with some accompanying text in three rows. With a large heading at the top, the starting point is clear. The charts switch from left to right and back to keep it lively.

It is tempting to do more with the leaf image, but I want to minimize clutter and keep it away from plot regions. It certainly shouldn't be used as a backdrop, however faded. The times of year could be augmented with holly for Christmas, leaves and snowflakes, but this would be pure chartjunk.

I had to restrain myself from commenting too much. I could have mentioned switches of governing political party, and providing details of the equipment used to keep leaves off the rails, but in this setting, I decided that was not my remit. You might feel that even mentioning the government subsidy was irrelevant. I found news stories online that explained many of the peaks in the line chart, but decided that it would overload the visualization to annotate it with them or have newspaper clippings. Different visualization jobs might call for a lot more information like that, and user testing can help to get the balance right. Remember to serve the reader and not your own curiosity or design ideas.

The Helvetica typeface, especially in the bold heading, is reminiscent of railway signs (at least in the UK). A little contextual design like this can get readers' interest, as long as it is done with a light touch and does not intrude. I thought about making it look like a railway station display board, but that is clearly going to favor style over substance.

Finally, the website of the Office of the Rail Regulator is cited as the source. You should always cite the data source. It is not too difficult to find exactly these data there, but if a lot of work had been done to combine and process data files before visualizing, then it would be more helpful to provide a link to the processed data files instead. (You can obtain my cleaned data file from the book's website at robertgrantstats.co.uk/dataviz-book.)

In summary, I had to consider layering information, color co-

Leaves on the line

Do Autumn leaves delay trains in SE England?

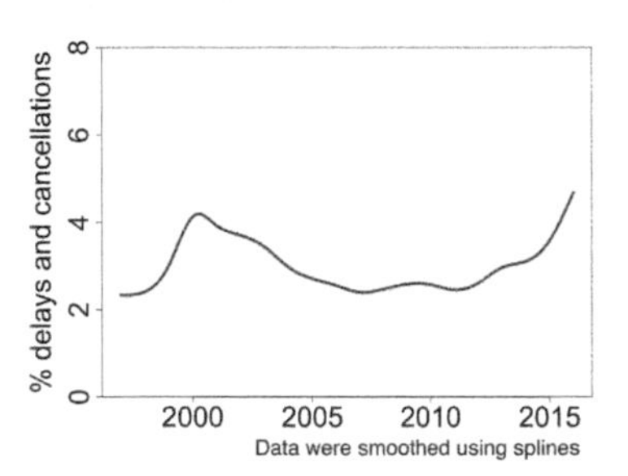

Delays and cancellations are recorded in 4-week periods.

The long-term trend was that performance improved, then deteriorated again.

Government subsidy peaked in 2006, and was reduced each year since then.

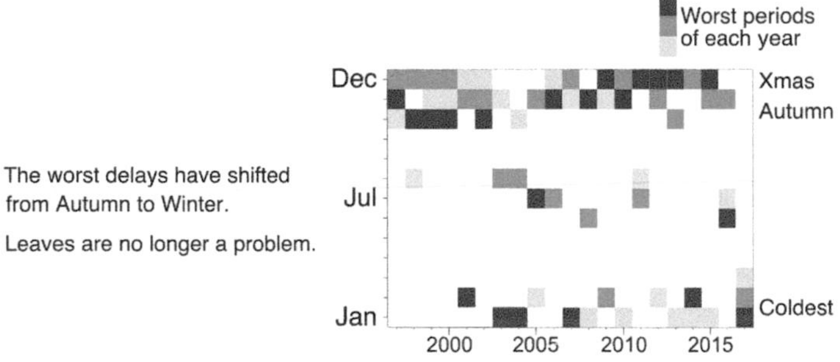

Short-term problems (spikes below) in context

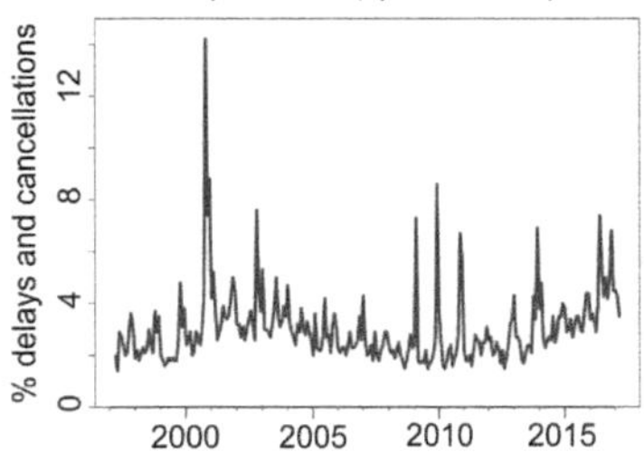

1994-97: Privatisation
1999: Paddington crash
2000: Hatfield crash
2000: Potters Bar crash; speed limits
2002: New infrastructure company
2009-11: Unusually snowy winters
2017: Southern Rail strikes

Data from dataportal.orr.gov.uk

Figure 16.2 The train delay data visualizations could be combined in a poster like this

ordination, avoiding clutter, and how much annotation or context to add.

16.3 TALK ABOUT IT

I mentioned user testing at various points in this book. When a lot of information is brought together into one package, this is especially important. Consider who needs to be involved to give that feedback and to make sure the desired impact is achieved. If time and resources permit, it may be possible to measure the understanding and recall of information among readers. Qualitative researchers can help here too, by interviewing and debriefing readers.

The goal of a good package including data visualization is to enable readers to access information in ways (and times and places) that they could not otherwise. In recent years, much has been written on the idea of the quantified self: that by using technology to track things like our physical activity and heart rates, we notice patterns that we are normally oblivious to, and can act on them to improve our lives (or at least, our efficiency at work). Data visualization is central to this; nobody experiences this kind of insight into themselves after downloading a large spreadsheet file of their personal data.

In contrast, designers Stefanie Posavec and Georgia Lupi, who became famous for their project Dear Data, stress the value of data humanism: using data to enrich our lives and not to drive efficiency at the cost of happiness. Somewhere on this spectrum you will find your readers' ultimate goals. It may be hard to elicit this sort of deep philosophical self-examination when they thought they were just hiring you to make a bar chart, but it can lead to a much more satisfying end result for all concerned if you pitch the information in the right way.

It's important to identify the level of statistical literacy in the intended readership, and make visualizations and text that is appropriate for them. Annotations will help to guide them through (Figure 16.3). There is a school of thought that says that a good visualization should be so intuitive that it requires no explanation – not even a legend or title – but in my experience this is a theoretical ideal that should not be allowed to get in the way of readers accessing and enjoying a visual interface to data. When we looked at Figure 6.1, I considered ways of annotating it with contextual

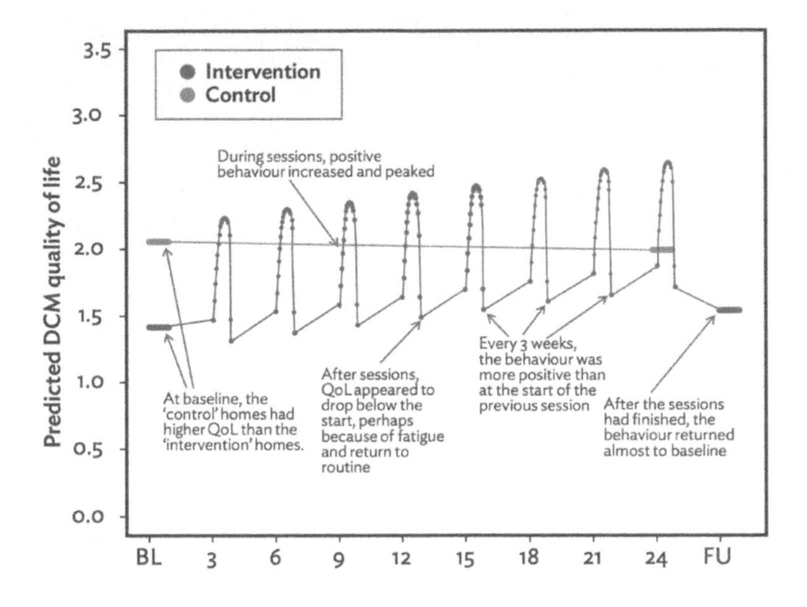

Figure 16.3 Changes in quality of life for nursing home residents with dementia as they take part in reminiscence sessions in a research project run by Royal Holloway, University of London. This line chart shows the results of a regression model, with annotations written in consultation with the academics and funders. The audience were health and social care professionals with relatively poor statistical literacy.

statistics from outside the study. This sort of context is always worth considering and discussing.

When visualizations are made by people with a statistics or data science background, they tend to use a lot of the classic formats and encodings and may be too reliant on bar charts, scatter plots and line charts. These are classics precisely because they are effective, but in a package, more engagement and recall can be achieved by bringing in some design thinking.

The context of the topic may give some direction to this. For example, if we are presenting data on commuting to work in Atlanta, Georgia (Chapter 3 and Figures 1.6 to 1.10), we could make one of the images a map with concentric rings at different distances, showing the number traveling a distance in each ring. This is just a

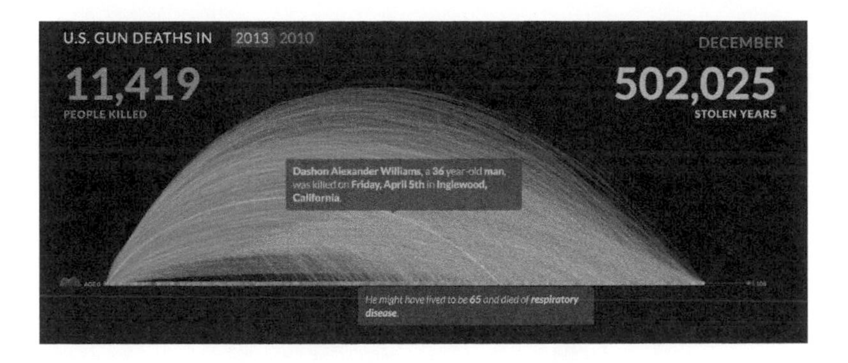

Figure 16.4 Gun deaths in the United States in 2013, an interactive visualization that appeals to the metaphors of flying objects in huge numbers. The reader can hover over the lines to see each one as a real person (with data taken from coroners' reports), then break down the numbers by age, sex, race and so on. Copyright Periscopic, reproduced with permission.

fancy histogram or bar chart, but it taps into the context: readers recognize and relate to the map, and if they are local, they will identify which ring they belong to.

Data visualization and design professor Isabelle Meirelles suggests that metaphor can be used in this way to make data seem more concrete and tangible to readers who are less statistically literate. This has to be tailored to each particular visualization and its context, but here are some closing examples that might inspire you.

Space flights are shown as concentric circles, suggesting orbits, in a poster created by 5W Infographics for *National Geographic* (you can view this at 5wgraphics.com). Lives cut short are shown as interrupted trajectories (Figure 16.4) in an online animation followed by interactivity by web agency Periscopic. The considerable depth of the ocean is shown in an extremely tall and thin graphic by the *Washington Post* that the reader has to descend (you can view this at goo.gl/hdKwBT). Casualties over time in Iraq are shown as a bar chart by Simon Scarr for the *South China Morning Post*, but flipping it upside down and using a gory red color suggests dripping blood (simonscarr.com/iraqs-bloody-toll). In one of my favorite visualizations, the volume of carbon emissions is shown as an ominous heap (Figure 16.5) in a video by Carbon Visuals.

Figure 16.5 Volume of carbon dioxide emitted in New York City in one hour, visualized as a heap of one-ton balls smothering Midtown. A still from a video at youtu.be/DtqSIplGXOA, by Carbon Visuals (www.realworldvisuals.com), reproduced with permission.

IV

Closing remarks

Further reading

THERE ARE SEVERAL GOOD BOOKS on data visualization, written in recent years, from a design or journalism angle. I particularly recommend:

- Alberto Cairo's *The Functional Art* and *The Truthful Art*
- Andy Kirk's *Data Visualization: A Successful Design Process* and *Data Visualisation: A Handbook for Data Driven Design*
- Isabelle Meirelles' *Design for Information*

For a more specifically business-oriented audience, Cole Knaflic's *Storytelling with Data: A Data Visualization Guide for Business Professionals* is a gently-paced introduction, aimed particularly at spreadsheet users.

William Cleveland's books, *The Elements of Graphing Data* and *Visualizing Data*, are an excellent introduction to the field from a statistical perspective, but the world has moved on in many ways since they were published in 1985 and 1993.

To think about the conflicting reasons for visualizing, and the compromises required in creating these images, you may be interested in:

- Gelman and Unwin's paper "Infovis and Statistical Graphics: Different Goals, Different Looks" is available at https://goo.gl/ij1ZaR and there are comments and responses archived at https://goo.gl/3jVSe4
- "Useful Junk? The Effects of Visual Embellishment on Comprehension and Memorability of Charts", by Scott Bateman and colleagues, is available at https://goo.gl/m9o4by
- Edward Tufte and David McCandless occupy opposite ends of a spectrum of visualization, from austere simplicity to colorful fun. They both have websites and many admirers, so you can find out about them before investing in any books. You may well decide that you agree with neither of them!

Nate Silver's *The Signal and The Noise* is an overview of the current state of data science, written for a general audience (so there's little mathematics). Pedro Domingos' *The Master Algorithm* talks about this from a machine learning and artificial intelligence perspective. If you have studied mathematics to the end of high school or first year of university, then you will get a much deeper and more practical understanding of the methods I've talked about in this book from *Computer Age Statistical Inference* by Brad Efron and Trevor Hastie.

Twitter has been a huge source of ideas, inspiration, and networking for me and many others in the world of #dataviz. I highly recommend using it as a way of quickly finding out what's trendy and who's doing what.

Blogs come and go, but I would particularly recommend Andy Kirk's at visualisingdata.com/blog. He regularly provides a long list of links to projects that he has been impressed by, so it is an excellent starting point for online exploration.

For inspiration, you might enjoy these more creative collections of dataviz:

- *Information Graphics*, by Sandra Rendgen and Julius Wiedemann
- *London: The Information Capital: 100 Maps and Graphics That Will Change How You View the City*, by James Cheshire and Oliver Uberti
- *Photoviz: Visualizing Information through Photography*, edited by Nicholas Felton
- *Information is Beautiful* and *Knowledge is Beautiful*, by David McCandless
- *Dear Data*, by Stefanie Posavec and Georgia Lupi

CHAPTER 1

- Alberto Cairo posted the Datasaurus at https://goo.gl/jiV7mh and an even more clever version appeared in this blog post: https://goo.gl/AFwQLa
- DrawMyData can be found at https://goo.gl/koeyCu
- Anscombe's quartet is described in the Wikipedia article at https://goo.gl/NgKFqM

- Gelman and Unwin's paper "Infovis and Statistical Graphics: Different Goals, Different Looks" is available at https://goo.gl/ij1ZaR and there are comments and responses archived at https://goo.gl/3jVSe4
- The election results visualizations were discussed in *Significance* magazine and this article is available online at https://goo.gl/usuRmf
- The background imagery of Figures 1.6–1.9 came from Google Maps, while that of Figure 1.10 came from Mapbox (mapbox.com) and was compiled by OpenStreetMap contributors.
- Designer Mike Monteiro gave a great talk on professional practice as a designer, available at vimeo.com/121082134 (there's some strong language). If you are considering working in dataviz, I strongly recommend listening and mentally substituting design for data visualization.
- Data design agency Periscopic wrote this excellent article on working with a dataviz consultant: https://goo.gl/x8mUBb

CHAPTER 2

- The train delay data came from dataportal.orr.gov.uk and I tidied them up, combining some different tables into one. You can also download my cleaned data in CSV format at robertgrantstats.co.uk/data/traindelays.csv
- Robert Kosara has compiled evidence about perception of data visualization, much of which is summarized in this talk: eagereyes.org/talk/how-do-we-know-that
- Isotype is explained in Robert Kosara's blog post https://goo.gl/x1cH2m
- The *Commuter Toolkit*, including photographs of 200 commuters in Seattle, is explained at https://goo.gl/AqTdTE
- Mike Kelley's photography: mpkelley.com

CHAPTER 3

- *Scientific American* had a blog post on Hal Craft and his astronomical data: https://goo.gl/f4ExrN

CHAPTER 4

- "The Global Epidemiology and Contribution of Cannabis Use and Dependence to the Global Burden of Disease: Results from the GBD 2010 Study", by Louisa Degenhardt and colleagues, can be read at https://goo.gl/kyXmTL – you will find that their visualization is rather different from mine!

CHAPTER 5

- A short blog post at r-bloggers.com/thats-smooth describes various smoothing algorithms succinctly.
- Spike histograms are available from the Hmisc package in R. An example is presented in Frank Harrell's tweet at https://goo.gl/WMgkw2
- The garden of forking paths is described in this magazine article: https://goo.gl/bcED27

CHAPTER 6

- I wrote more on odds ratios and relative risks in a paper called "Converting Odds Ratios to a Range of Plausible Relative Risks for Better Communication of Research Findings", which you can access at https://goo.gl/ahQQVN
- The alcohol and cancer paper from the Million Women Study is at https://goo.gl/Zi8YmN. The press release is at https://goo.gl/R2d2ig and the comment article is titled "Action Needed to Tackle a Global Drink Problem", by Ian Gilmore, and published in *The Lancet*, 27 June, 2009.

CHAPTER 7

- Much of the research about perception of different visual parameters is summarized in William Cleveland's books, or Isabelle Meirelles'.
- Robert Kosara has compiled evidence about perception of data visualization, much of which is summarized in this talk: eagereyes.org/talk/how-do-we-know-that
- The bird feeder chart is updated periodically with measurements from my garden and located at https://goo.gl/XWbhrn

- John Tukey's book *Exploratory Data Analysis* is now out of print and expensive second-hand, but contains a lot of visualisation, much of it experimental and focused on quickly "scratching down" on paper to get an impression of data. In the statistics literature, it is uniquely humorous and iconoclastic.
- There is a short biography of Stanley Smith Stevens at https://goo.gl/TmgndU
- A good place to start learning more about optical illusions is the website of Professor Akiyoshi Kitaoka at https://goo.gl/TcpzBB

CHAPTER 8

- The bootstrap and related randomization procedures are the main approach to teaching statistical inference in the Locks' textbook, *Statistics: Unlocking the Power of Data.* There are many other introductions, but all requiring some comfort with reading algebra and probability notation. Using simulation to introduce inference is now recommended by the American Statistical Association in their *Guidelines for Assessment and Instruction in Statistics Education* (GAISE).
- Bias comes in many forms, and most are clearly explained in a paper called "Bias" by Miguel Delgado-Rodríguez and Javier Llorca, available at https://goo.gl/HHxY4F
- Sir David Spiegelhalter's blog on funnel plots is a good place to start exploring these further: understandinguncertainty.org/fertility
- To read more about funnel plots, performance indicators, and public service management, I recommend the book, *Performance Measurement for Health System Improvement*, edited by Peter Smith and colleagues, and published by Cambridge University Press.

CHAPTER 9

- Edward Tufte's online writing on sparklines is collected at https://goo.gl/onjZtd

- There is a short explanation of how to read ternary plots (for geologists, but no knowledge of the subject is needed) at https://goo.gl/MEi7QX
- Hans Rosling presented a great example of his animated bubble charts for the BBC, which you can watch at https://goo.gl/gy7ucd
- The paper, "The Heterogeneous Dynamics of Economic Complexity", by Matthieu Cristelli and colleagues, is available at https://goo.gl/3zr4Ct
- Andrew Elliott's design work is hosted at andrewelliott.design
- My stop-frame animation programs for R and Stata are available at https://goo.gl/Z5z5eQ
- The thinkpurpose blog post is available at https://goo.gl/ohEbQZ

CHAPTER 10

- If you want to know more about biomedical applications of data science, an excellent starting point is Doug Altman's book, *Practical Statistics for Medical Research.*
- The IMPACT study was published in scientific papers by Jane McCusker and colleagues, the most relevant of which is "The Effects of Planned Duration of Residential Drug Abuse Treatment on Recovery and HIV Risk Behavior," available at https://goo.gl/87SrdE
- The concept of marginal effects is explained in more detail here: https://goo.gl/tDKDRr

CHAPTER 11

- The TensorFlow Playground is located at: https://goo.gl/U7vWqV
- For those with a little mathematical training, the neural networks chapter in Efron and Hastie's book, "Computer Age Statistical Inference: Algorithms, Evidence, and Data Science" is excellent.

- "Interpretable Explanations of Black Boxes by Meaningful Perturbation," by Ruth Fong and Andrea Vedaldi, is located at: https://goo.gl/CcJokD.
- Another relevant effort to visualize CNNs is "Picasso: A Free Open-Source Visualizer for Convolutional Neural Networks," by Ryan Henderson: https://goo.gl/mJgHzX

CHAPTER 12

- Principal components analysis, one of the most commonly used dimension reduction methods, is explained in more detail and without excessive mathematics at https://goo.gl/rXR2SE ; other methods such as correspondence analysis and multidimensional scaling operate on the same general principle of projecting and rotating the data.
- "How to Use t-SNE Effectively" is a good starting point to understand this more complex procedure in more detail: https://goo.gl/r9iTWF
- Joel Caldwell's blog post on clustering is at: https://goo.gl/PSHNty

CHAPTER 13

- Parag Khanna's book, *Connectography*, has many examples of network analysis and visualization.
- Maarten Lambrechts blogged about dot density maps at https://goo.gl/c7yjuE
- Naomi Robbins' book, *Creating More Effective Graphs*, introduces minimaps.
- *Long Barrows in Hampshire and the Isle of Wight* was published as a report of the Royal Commission on Historical Monuments in 1979, and contains several fine diagrams and maps. The copyright now belongs to English Heritage.
- Robin Wilson's digitized cholera data are available at blog.rtwilson.com/john-snows-cholera-data-in-more-formats and you can read about John Snow and his map at https://goo.gl/hDYPCu

- Alec Rajeev's cartogram is online at https://goo.gl/WGL5xa
- The United Nations High Commissioner for Refugees Operational Data Portal is at https://goo.gl/HLN3HS
- Diego Valle-Jones' contour map of Monterey homicides is online at bl.ocks.org/diegovalle/5166482
- Chris Whong's "NYC Taxis: A Day in the Life" is online at https://goo.gl/CL1fQo
- "Longterm Trends in the Contribution of Major Causes of Death to the Black-White Life Expectancy Gap by US State," by Corinne Riddell and colleagues, is available online at https://goo.gl/iyH29D
- You can read more about hurricane maps and misconceptions about uncertainty at https://goo.gl/oVuVBW
- Sophie Engle's edge bundling map of flights is at https://goo.gl/spDWnM and upon loading, you will see the edges being progressively bundled.

CHAPTER 14

- To learn more about interactive graphics, the best thing to do is to look out for online articles with interactivity and consider what interactions they use, what you like and what you don't like. The best publications for this, in my opinion, are *The Economist*, the *Financial Times*, the *New York Times*, the *Washington Post*, and *The Guardian*. However, there is a lot of competition in this market, so look around and judge for yourself; this list will soon be out of date.
- If you want to make interactive content yourself, you will need to invest some time in learning either JavaScript (alongside HTML and CSS to construct a web page) or a software tool that does the work for you. For all the essentials of web coding, I suggest w3schools.com, which is free online. For a thorough but accessible introduction to JavaScript, *Eloquent JavaScript: A Modern Introduction to Programming*, by Marijn Haverbeke, is highly recommended (and also free online at eloquentjavascript.net).

- The D3 JavaScript library is well introduced by Scott Murray in his book, *Interactive Data Visualization for the Web*, and also in *D3 Tips and Tricks* by Malcolm Maclean. The official website is d3js.org

- For the Leaflet library, I recommend starting with the tutorials on the website at leafletjs.com, then moving on to Paul Crickard's book, *Leaflet.js Essentials.*

- Mapbox (mapbox.com) is also useful for quickly generating a JavaScript interactive map, which you can then customize further once you have some familiarity with the language.

- If you want to look into responsiveness for a web page, start with JavaScript libraries like Bootstrap or Flexbox.

- To learn more about the SVG graphics format, I highly recommend the works (books, blogs, videos) of Sarah Drasner and Nadieh Bremer. You can find them easily online.

- The examples included in this chapter are:

  - State of Obesity: stateofobesity.org
  - *The Guardian* article "Bussed Out: How America Moves Its Homeless": https://goo.gl/MKUV23
  - The *New York Times* article "How the Recession Reshaped the Economy, in 255 Charts": nyti.ms/2jVJvTM
  - Amanda Cox for the *New York Times* in 2010 with Flash: goo.gl/RtUPB2
  - TensorFlow Playground: playground.tensorflow.org
  - Paul Lambert's splines: https://goo.gl/d5AvPV
  - How to use t-SNE effectively: https://goo.gl/r9iTWF
  - Bayesian Estimation Supersedes the t-test: https://goo.gl/2rcx5p
  - StatKey: lock5stat.com/statkey

CHAPTER 15

- *FOILing NYC's Taxi Trip Data*: https://goo.gl/uDVFqZ (also see *Analyzing 1.1 Billion NYC Taxi and Uber Trips, with a Vengeance*, by Todd W. Schneider, at https://goo.gl/pid8s8

- *Bin-Summarise-Smooth: A Framework for Visualising Large Data*: https://goo.gl/KGmhjX
- For open-source software tools that are designed for big and fast data, look up the ever-evolving set of software made by the Apache Foundation (projects.apache.org)
- Oliver O'Brien's Bike Map: bikes.oobrien.com
- Lightning Maps: lightningmaps.org
- Flightradar24: flightradar24.com

CHAPTER 16

- Simon Sinek explains this idea of why, how, what in a short TED talk at https://goo.gl/Vwv6Ke
- Noah Iliinsky's views are expressed in the book *Beautiful Visualization: Looking at Data through the Eyes of Experts*, by Julie Steele and Noah Iliinsky.
- The proponents of Quantified Self collect examples of data visualisation at https://goo.gl/bExtd9
- Stefanie Posavec and Georgia Lupi's project *Dear Data* is online at dear-data.com, and published as a book. Georgia discusses data humanism at https://goo.gl/gZpTm7
- The RADIQL project report, on reminiscence therapy in dementia, is at https://goo.gl/b2gdWC
- *Gun deaths in the US*: guns.periscopic.com
- Fifty years of exploration (the solar system): https://goo.gl/2nRKnt
- The *Washington Post* article "The Depth of the Problem": https://goo.gl/hdKwBT
- *Iraq's Bloody Toll*, by Simon Scarr: https://goo.gl/pCVSJy
- The animated video *New York City's greenhouse gas emissions as one-ton spheres of carbon dioxide gas* is at: youtu.be/DtqSIplGXOA

Index